SPIDERS

The Ultimate Predators

SPiDERS

The Ultimate Predators

Stephen Dalton

A & C BLACK • LONDON

Published in Great Britain in 2008 by
A&C Black Publishers Limited
38 Soho Square
London W1D 3HB
www.acblack.com

Published by arrangement with
Firefly Books Ltd.
66 Leek Crescent
Richmond Hill, Ontario
L4B 1H1
Canada
www.fireflybooks.com

ISBN 978-1-408-10697-6

Interior design by Kathe Gray Design

Printed in China

10 9 8 7 6 5 4 3 2 1

Contents

Introduction

If you wish to live and thrive
Let the spider run alive

— *traditional English rhyme*

Wherever we are, there is likely to be at least one spider within a few feet of us. It may be weaving a web, floating by on a gossamer thread, dancing to its mate, sucking the juices from an anesthetized fly or simply resting under our chair. Whatever it is doing, we are privileged to share the planet with about 40,000 known species of these remarkable animals.

Spiders are the most successful terrestrial predators on earth, occupying almost every niche possible. They are found from mountaintops to seashores and from ponds to deserts. They can even be found thousands of feet up, traveling vast distances as they balloon through the air on threads of gossamer. Spiders have been around for some 400 million years now and were pivotal in the evolution of insects, the most abundant class of animals on earth. Spiders consume more prey than all other carnivores. The late W.S. Bristowe, a British arachnologist, established that at certain times undisturbed meadows can support an astonishing population of more than two million spiders per acre, and that the weight of insects consumed per year by spiders easily exceeds the weight of the entire human population of England.

The success of spiders is almost wholly due to the formidable and astonishing array of techniques they have evolved for trapping insects and other small creatures. Their tactics are the result of a 300-million-year arms race fought against the insects. These range from a variety of ingenious traps in the form of webs to a host of other devious methods, including lassoing, jumping, stealing, chasing, ambushing, spitting, fishing, masquerading as other animals and even attracting prey by

◄ Webs on an autumn morning.

European house spider (*Tegenaria domestica*) using its chelicerae, or jaws, to clean its foot.

mimicking the prey's pheromones. My fascination with this stunning diversity of hunting techniques largely inspired the creation of this book.

It seems unfortunate that many naturalists and organizations concerned with conservation have tended to concentrate on the more obviously attractive groups of invertebrates, such as butterflies and dragonflies, as is clear from the many books on those subjects. The vast majority of spiders are not brightly patterned and colored; being tender and vulnerable creatures and surrounded by a profusion of enemies, they rely on merging with the subtle shades of their surroundings. Their coloring is usually composed of beautifully delicate patterns of browns, greens and grays, as reflected in the photographs in this book.

Another reason for their comparative neglect is that spiders are often tricky to positively identify. More than half are only 0.04 to 0.2 inch (1–5 mm) long, and the differences between many species can be discerned only under a microscope. Here, though, we are concerned mainly with the larger and more significant species. The difficulties of spider identification are hardly made easier by their numbers—more than 640 named species in England alone and about 3,700 in North America. In contrast, there are only about 65 British butterflies and around 700 in North America.

Spiders also have an image problem, so this book attempts to offset their creepy reputation. They possess rather a lot of legs to worry about; they are known to have a poisonous bite; they sneak about in dark places, scuttle across the floor at high speed and leave untidy webs all over the place. Paradoxically, though, many spiders are actually creatures of sunlight and are not at all creepy. The jumping spiders, by

Pantropical jumping spider, from a family of sunshine lovers.

far the most numerous single group, have an enchanting and, some would say, almost cuddly appearance as they run jerkily over rocks and tree trunks, their large eyes following our every movement.

The aspect that most concerns many of us is actually the last thing we need to worry about—their bite (apart from some notorious exotic species). Even the few non-exotics that can nip do so only in self-defense, when severely provoked or squashed against our body. In reality most spiders are very nervous and retiring creatures because, unlike insects, which are protected by a tough exoskeleton, spiders have soft and vulnerable bodies. They do everything possible to keep out of harm's way, disappearing into their hiding places at the slightest disturbance or sign of danger. Many come out into the open only at night, spending the day hiding in some crevice or curled up in a leaf.

Spiders, like so many other animals, including humans, are predatory carnivores, but they are more humane than most others, albeit unconsciously so. This might seem like anthropomorphizing, but many of the predators we admire—such as owls, eagles and tigers—usually tear their relatively intelligent victims limb from limb while they are still alive, while the spider will first anesthetize its prey or, more likely, kill it with a lethal injection! We should also bear in mind that the nervous system of a spider's invertebrate prey is thousands of times less elaborate than that of any vertebrate, so their capacity to suffer is insignificant compared with the prey of larger carnivores.

There are a number of ways of organizing a book of this nature; for example, by family or habitat. Most spider books are based on classification, which is a sensible approach for identification and more serious study of the subject, but a slightly different course has been adopted here. In view of the range of ingenious hunting methods used by spiders, in this book they are grouped as follows: those that hunt down their prey by chasing, those that lie in wait and ambush, spiders that leap onto their victims, and of course the majority, which spin webs to trap insects, the latter being subdivided by different types of webs—orb webs, sections of orbs, sheet webs and funnel traps. Finally there are spiders that don't neatly fit into any of these categories—the nonconformists, those that tend to employ even more freakish techniques such as spitting, fishing and raiding other spiders.

It will soon become clear that these demarcations are not set in stone, as some species within one group often share the characteristics of another. For instance, many spiders that are capable of chasing down their prey may sit and wait for prey to come within reach before making a quick dash, so these could equally be described as ambushers. Similarly, some of the web builders do not actually trap prey in their webs but dash out of a hole or tunnel at high speed as soon they sense an insect touching a strand; and there is a good case for classifying fishing spiders with the chasers instead of bundling them with the nonconformists. Nevertheless, the broad divisions adopted here do help to demonstrate the range of hunting styles employed by these astonishing animals.

Unless it is unusually spectacular, I have avoided explaining much about courtship and mating behavior, particularly as several other books, including W.S. Bristowe's *The World of Spiders,* cover this topic in lavish detail. What has been included is a rough-and-ready translation of the scientific names. Whereas most of the names appear sensible and logical, others do not seem to have any obvious connection with the spider's appearance or way of life. Nevertheless I find them interesting—and at times amusing.

This book would not be complete without mentioning its omissions. For instance, we hardly touch upon the many minute spiders that require high magnification to see, let alone identify. These spiders mostly belong to a single vast family, the Linyphiidae; two of the larger species are portrayed here. Also excluded are many families specific to the tropics and Australasia and Africa. What is covered are the most important families common to both northern Europe and North America, together with a few representative "special" spiders that thrive on one continent but not the other. For example, the ladybird spider (*Eresus*) and the water spider (*Argyroneta*) are both European species that are not found in North America, while the black widow (*Latrodectus*), the golden orb spider (*Nephila*) and tarantulas do not enrich the lives of English country folk (perhaps global warming may change this!).

It may prove surprising to learn that there is considerable overlap between the actual species: over 30 percent of those described in this book are common to both continents, while a larger number of others are very similar. Many species have been introduced by shipments of plants, furniture and other imports from Europe, and a few species such as the European house spider (*Tegenaria*) and the daddy longlegs spider (*Pholcus*) have worldwide distribution. The species that are very similar have for the most part evolved over the eons from the same sources as temperate Eurasian fauna.

➤ A garden spider, the archetypal European spider that is also found in North America, in its freshly made symmetrical web.

Crab spider (*Misumena vatia*) showing the crescent-shaped eye arrangement characteristic of the family.

I am often asked, "What is the point of spiders?" I am tempted to reply, "What is the point of humans, or anything for that matter?" Does there have to be a point or reason for the existence of a particular species? Life has evolved as a result of the universal process of natural selection, just as everything has since the big bang. What is certain is that all plants and animals rely on one another for their continued existence. In fact, the small and "lowly" creatures on which larger animals depend are more crucial in the scheme of things than lions and pandas. Over the eons, evolution ensured that living things remained generally in balance with their environment and with each other—until, that is, man began to overpopulate the planet and destroy this balance. Now, in a relative twinkling of an eye, nature as we used to know it is ending.

So what specific roles have spiders played? It is estimated that, worldwide, spiders are largely responsible for about 99 percent of the insect mortality rate—although spiders are incapable of discriminating between the so-called harmful and beneficial insects. Trials have proved that spiders can control large numbers of insect pests in agricultural areas, but since spiders have been so poorly studied their explicit function in nature has not been fully demonstrated. Spiders are food for a huge variety of animals such as birds, mammals and fish, while their silk is used in nest construction by many birds. Clearly, by virtue of their enormous population, the effect spiders have on maintaining the balance of nature is enormous.

Compared with studies of other, diverse terrestrial groups of animals, spider surveys are relatively simple to conduct, as few time-consuming dissections are required for their identification. This fact, together with their wide range of sizes—0.016 to 4.8 inches (0.4–120 mm)—and as much range in their biology as their size, now causes many ecologists to believe that spiders are the ideal subjects for assessing habitats and biodiversity, because they more easily provide information about the value of habitats than higher plants or vertebrates.

Economics and benefit to mankind seem to be the only language that many of us understand, so in addition to their importance as insect predators, we need also to take account of spiders' ability to produce silk and venom. Researchers are still struggling to work out how to take advantage of the combined strength and elasticity of spider silk so that it can be mass-produced. Venoms too are being studied for their potential for treating pain, epilepsy, strokes and Alzheimer's disease.

Perhaps we should also consider the spiritual and aesthetic value of these animals. Few of us, including arachnophobes, can fail to wonder at a dew-laden web on an autumn morning or to admire the spider's subtlety of form and color. Being moved and absorbed by the lives of spiders and all living things, however small and superficially insignificant they may appear, helps us to value all life as well as our own and realize that everything is intimately connected to everything else. Spider watching isn't only about spiders; it's about the whole natural world. In my view if such a holistic approach was embraced by everyone the world over, survival of Earth's natural beauty and life as we know it would be guaranteed.

According to E.O. Wilson, the farsighted American ecologist, we have a genetically inbuilt love for wild places and living things—he calls it biophilia—although for many of us this affinity is being increasingly suppressed by our material and anthropocentric approach to modern life. As Wilson says, humanity is exalted not because we are far above other living creatures, but because knowing them well elevates the very concept of life.

I quote a prophetic article from the *Illustrated London News* written more than a hundred years ago: "Man cannot wait for the cooling of the earth before consuming everything in it from teak trees and hummingbirds to snakes and spiders. In a hundred or two years hence he will be perplexed by a world in which there is nothing except what he has made." *Perplexed,* I think, is an understatement. We share the world with tigers, whales, hummingbirds and, yes, spiders too. The world would be a poorer place without any one of them.

1

What is a Spider?

Before delving into the spiders themselves, it should prove helpful to give some general background information about these animals—how they differ from other invertebrates, how they are classified, how they build webs and mate.

Structure

The most obvious characteristic of a spider is its eight legs. Insects have only six. Unlike insects, whose adult bodies are divided into three parts, the spider's has only two parts. Also unlike insects, there is no larval or pupal stage—apart from color, pattern and reproductive organs, the young generally look like miniature versions of their parents immediately after hatching. The eggs are protected in silken egg sacs that come in a wide variety of sizes and designs. As the spiderlings become older, they shed their exoskeletons to accommodate their increasing size. The number of moults depends largely on the size of the spider; the smaller species undergo two or three while the larger ones moult up to about a dozen times. Lost or damaged limbs are often regenerated during or after the moult.

In contrast to their relatives, the mites, harvestmen and scorpions, which have their heads fused with the body, the two parts of a spider's body are connected by a very narrow stalk, although this is normally out of sight. Spiders also boast a unique feature—spinnerets. These are located at the tip of the abdomen for the all-important task of silk production.

◄ Cucumber or green orb weaver spider.

Harvestman (*Mitostoma chrysomelas*).

Scorpion from the Kalahari.

Cephalothorax

The front part of the body, the cephalothorax, consists of a fused head and thorax that is protected by a hard chitinous layer called the carapace. The front end accommodates the eyes—the vast majority of spiders have eight, although there are three families that possess only six (Oonopidae, Dysderidae and Scytodidae). The shape of the spider's face, the variety, number and arrangement of the eyes, and the shape of the jaws are key features in distinguishing one family from another. The eye color also can vary from a pearly luster to dark, but it can sometimes be difficult to see because of the presence of hairs.

Abdomen

The abdomen varies considerably in shape, markings and size. Even within a single species the size can vary enormously, depending on feeding or the stage of any eggs developing inside. The upper side frequently has a cardiac mark and four depressed brownish spots called sigella, which mark the internal muscle attachments. At the top of the underside are two pale patches (four in certain groups) that mark the position of the book lungs—air-filled cavities containing blood-filled leaves.

The tip of the abdomen holds the spinning organs, or spinnerets. Primitive spiders have eight spinnerets, but most present-day species appear to have lost the front two completely. In some the front two spinnerets have evolved

Spider eyes

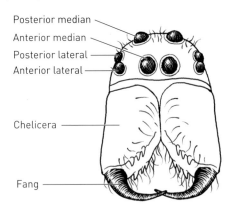

Posterior median
Anterior median
Posterior lateral
Anterior lateral

Chelicera

Fang

➤ Mouse spider showing ventral surface with book lung patches.

A scaffold-web spider (*Theridian pallens*) protecting its egg-sac.

into a single flat plate, the cribellum, which produces thick bands of bluish tangling silk. Just below the book lungs of most adult females may be seen the epigyne, the genital opening with a complexly shaped plate that is the lock to the male palpal key.

Jaws

The jaws, or chelicerae, are the spider's main weapon and are used for subduing prey. They are divided into a stout basal segment and an articulated thorn-like fang. The tip of the fang has a fine opening through which venom can be injected from a duct supplied by a gland in the cephalothorax. When not in use, the fang

Structure of Mouthparts

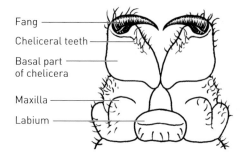

Fang —

Cheliceral teeth —

Basal part of chelicera —

Maxilla —

Labium —

External structure of a female spider

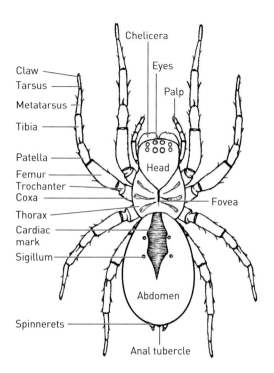

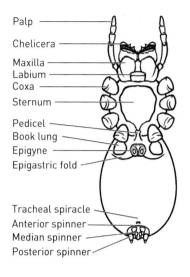

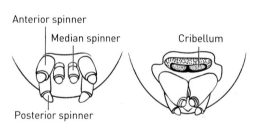

folds into a groove in the basal section, which is often bordered by teeth on the front and rear edges. The relative size of the chelicerae varies greatly between different species, which is also a useful feature in identification. Behind the chelicerae is the mouth, which is used for sucking up the liquid contents of prey.

Palps

Spiders have palpal organs at the front of the head that are really a fifth pair of small modified legs. These are important sensory organs and are used in prey manipulation. In the male the palps also perform an extraordinary role in reproduction. Male spiders are unique in the animal world because they pick up and carry sperm cells ready to be injected into the female. Moreover, the palps fit like complex keys into a lock-like plate on the belly of the female. The complexity and diversity of these organs can be fully appreciated only under the microscope. Because every species has its own special lock and key, the male palps and the female epigyne are vital characteristics for positive identification of most species.

Legs

The four pairs of legs are divided into several joints; the tip of the terminal segment, the tarsus, has claws. Web spiders generally have three claws that are used for web control (see page 25), while some wandering and hunting spiders have lost one claw, often replacing it with a tuft of hairs (scopulae) that give a better grip. Some families have a series of curved bristles called the calamistrum running along the dorsal edge of the metatarsus of the fourth pair of legs. This is used to comb out a viscid substance from the cribellum that, when combined with ordinary silk, produces a thick, lace-like, fluffy bluish web.

The legs are clothed in a wide variety of types of hairs and spines, each of which has a special role. Most have a sensory function and possess their own nerve supply at the base. Some hairs are sensitive to touch; some at the end of legs are chemosensitive, allowing taste by touch; while other, fine vertical hairs, the trichobothria, are highly sensitive to air currents and vibration. More robust hairs are spine-like and assist in prey capture. Jumping and lynx spiders often have flat hairs rather like the scales on butterfly wings, which reflect light to produce iridescent colors.

Other sensory organs capable of detecting pressure, humidity and heat may also be present on the legs or body in the form of hairs or slits.

◄ Golden orb spider (*Nephila clavipes*) moulting.

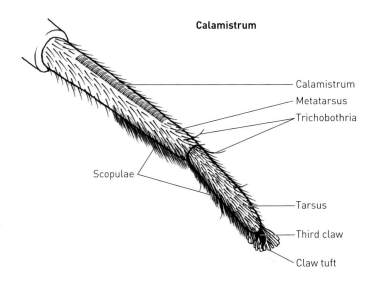

Calamistrum

- Calamistrum
- Metatarsus
- Trichobothria
Scopulae
- Tarsus
- Third claw
- Claw tuft

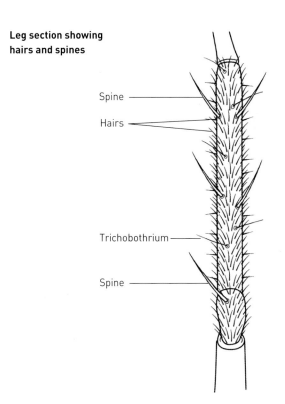

Leg section showing hairs and spines

Spine
Hairs
Trichobothrium
Spine

Spider names and classification

Together with other jointed-leg creatures such as crabs, scorpions, millipedes and insects, spiders are classified as arthropods. Unlike adult insects, which have six legs and usually wings, spiders and their kin, including mites and scorpions, possess eight legs and belong to the class Arachnida. Within Arachnida all spiders belong to the order Aranea (not to be confused with the Araneidae family of orb weavers; see page 30). On the basis of structure and behavior, spiders are further classified into two suborders and over a hundred families.

Earth supports some 40,000 named species of spiders but there may be three times as many yet to be discovered—if their habitats are not destroyed first. Even in a country as heavily populated with naturalists as Britain, species new to science are discovered regularly, adding to the 650-odd that have already been identified. In North America there are undoubtedly many species yet to be named.

Clearly some intelligent way of dividing spiders into groups with similar characteristics is vital if we hope to make any sense of such vast numbers. Very few species have been given English names—or, for that matter, names in other languages—and even if they had, the names would most likely be based on different characteristics anyway.

Internationally agreed-upon scientific names are vital if chaos is to be avoided. The name "house spider" could refer to any of several species—indeed, the American house spider belongs to an entirely different family from the European "house spider," while the name "jumping spider" could refer to any of 5,000 species! English-language family names can also give rise to much confusion. Take, for example, the Theriidae. This family has been

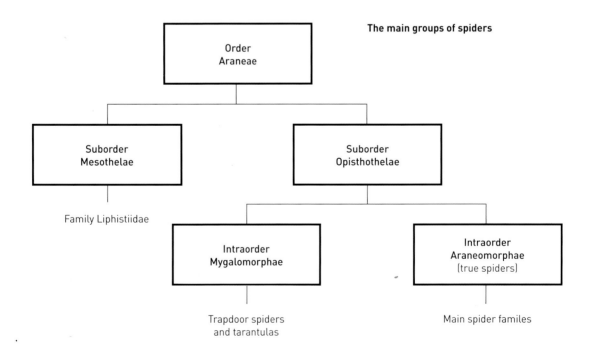

The main groups of spiders

Order
Araneae

Suborder
Mesothelae

Suborder
Opisthothelae

Family Liphistiidae

Intraorder
Mygalomorphae

Intraorder
Araneomorphae
(true spiders)

Trapdoor spiders
and tarantulas

Main spider familes

The classification and meaning of *Heliophanus favipes*

Kingdom	Animalia	as opposed to plant kingdom
Subkingdom	Metazoa	many-celled (as opposed to single-celled)
Phylum	Arthropoda	literally "jointed limbed"
Class	Arachnida	includes harvestmen, ticks, mites and spiders
Order	Aranea	spiders
Suborder	Araneomorphae	spiders with pinching jaws
Family	Salticidae	jumping spiders
Genus	*Heliophanus*	from Greek for "found in the sunshine"
Species	*flavipes*	red-legged

given at least four different names in the English language alone: scaffold-web spiders, space-web spiders, cobweb spiders and comb-footed spiders.

Scientific names are not devised willy-nilly but are closely linked with structural or other qualities that reflect the relationship between one group and another. Classification, or taxonomy, is simply a filing system for living organisms based on evolutionary relationships, taking the form of a branching hierarchy. At the top sits the kingdom, which is further divided down the tree into subkingdom, phylum, class, order, suborder, family, genus and finally species. Sometimes to complicate the matter (or to simplify things, depending on which way you look at it), extra subdivisions are added, such as subfamily or even subspecies.

By convention, family names end in *idae,* but they are often used adjectivally as well—for example, the family Araneidae may be referred to as araneids. In this book only the proper family name is capitalized.

The table above shows how a common little jumping spider, *Heliophanus flavipes,* is classified.

At this point it is worth saying something about the suborders into which spiders may be placed. It concerns the fundamental differences in structure between spiders from which the various families spring.

Suborder Mesothelae

The Mesothelae are a very ancient group of spiders from which all subsequent spiders have derived. There is only one family within the Mesothelae surviving today and it contains a small number of species; all are found in Southeast Asia, so they are not discussed in this book. The interesting feature about these spiders is that, like insects but unlike present-day spiders, they have a segmented abdomen. They also have eight spinnerets in the center of their abdomen. All live in caves or underground burrows covered by trapdoors.

Suborder Opisthothelae

All the spiders in this book are members of this suborder. They are further divided into two infraorders, the Mygalomorphae and the Araneomorphae.

Infraorder Mygalomorphae

Spiders in this group are familiar to most people as the huge, hairy bird-eating spiders, or tarantulas, but they bear no relationship to the true tarantulas from Spain, which belong to the wolf spider family, Lycosidae.

The mygalomorphs are a relatively primitive group consisting of 11 families and characterized by large, forward-projecting chelicerae that operate with a parallel upward and downward movement of the fangs. These spiders also have two pairs of book lungs as opposed to the single pair in "modern" spiders. Like the Mesothelae, the majority of mygalomorphs live in underground holes, often with a trapdoor entrance—hence their other name, trapdoor spiders. Whereas North America has several species, northern Europe has only one representative, the purse-web spider, *Atypus affinis,* although it is rare and localized in England (see page 174).

Infraorder Araneomorphae

The araneomorphs, sometimes called the true spiders, comprise the vast majority of spiders and are regarded as being more highly evolved. The difference between these spiders and those in the other two suborders is the way in which the chelicerae are attached to the head and their sideways action, which allows them a greater biting span. They are also capable of making many different kinds of silk and have evolved tracheae for breathing, having dispensed with one of the two pairs of book lungs of the mygalomorphs (insects have taken this further in having only tracheae).

The many families (more than 80) that form the Araneomorphae reflect the great diversity of body forms and lifestyles that has allowed this group to colonize all corners of the globe. Spiders from different families are usually clearly different in many ways.

Finally, within the many families of spiders there are one or more genera. The name of the genus, which always starts with a capital letter, forms the first part of the scientific name. Within each genus there may be one or more species, and the specific (species) name forms the second part of the scientific name; by convention this binomial name should always be printed in italics. Typically the species within a genus look fairly similar and tend to have a similar lifestyle, but they may be adapted to different habitats.

Because spiders have been rather neglected over the years compared with most other animal groups, and as new species are discovered, taxonomists frequently revise their ideas about the relationship between one spider and another. As a consequence scientific names are changed and species are sometimes transferred from one genus to another. So beware—the names used in one book frequently disagree with another! In the near future, though, DNA analysis may change our view of spider family relationships.

Showing orb web spider's use of claws for handling web.

Spider silk

Various invertebrates produce silk—certain insect larvae and mites, for example—but no animal approaches the spider in the versatility or variety of ingenious ways it is employed. The silvery strands of silk are woven into almost every aspect of a spider's life, and it is silk that is largely responsible for spiders' huge success over the past 300 million years, allowing them to compete with insects. Apart from making silken snares for catching prey, spiders can produce several different types of silk with a variety of characteristics from up to six different silk glands: sticky, viscid silk; silk for attaching threads; a sticky substance for depositing on thread; silk for wrapping prey; woolly cribellate silk; dragline silk; silk to protect egg-sacs; and "gossamer" silk suitable for ballooning tiny spiders thousands of feet up into the sky.

Silk is a fibrous protein made up of chains of amino acids produced by special glands in the abdomen. Although it appears as a single thread to the naked eye, in fact it is made up of several very fine strands of between one-millionth and four-millionths of an inch (0.00025–0.001 mm) in diameter. It starts as a liquid that is pushed through long ducts leading to microscopic spigots on the spider's spinnerets, but the extrusion process causes the liquid to solidify into strands. Most spiders have two or three pairs of spinnerets at the rear of the abdomen.

Triangle spider wrapping up old web prior to re-cycling it. The slight blurring of the legs is due to their rapid movement while rolling-up the silk (also see pages 135-137).

Valves on the spigots control the thickness and speed of the extrusion. As the spigots release the liquid out of the ducts into the air, its molecules are stretched out to form long strands that are finally wound into a strong silk fiber by the spinnerets. Watching spiders do this with the aid of a magnifier or through the camera lens is quite mind-blowing, for as well as manipulating the spinnerets or cribellum they can sometimes be seen to operate all eight legs simultaneously, each tarsal claw either working with the separate threads or holding the spider onto its web.

The whole operation is of course instinctive but it always reminds me of an organist playing a Bach fugue with ten fingers and two feet.

Each of the spider's several silk glands is optimized to produce a different quality of silk. Thus, by winding a range of varieties of silk together in varying proportions, the spider can form an assortment of webs. Furthermore, silk can be made into multiple layers, followed by a coating of a variety of substances suited for different purposes—sticky or waterproofing materials, for example.

The strength and elasticity of spider web is legendary, some types being five times stronger than steel of the same thickness, and it is capable of being stretched to about 10 times its original length. The secrets of its composition are still not totally understood to this day.

Silk is expensive in terms of body resources, so when the web is taken down after it becomes damaged or the spider wants to move to a fresh location, it is recycled. The spider does this by rolling up the silk into a ball and eating it as it climbs up or moves along the web.

Construction of a typical orb web

The most difficult part in the construction of an orb web is the first thread. This needs to be a sturdy horizontal line from which the rest of the web will hang. So how does the spider place this thread between the two connecting points? The answer is simple. It makes use of the wind and some luck: the wind carries away a thin silken thread from its spinnerets, and if the spider is lucky the thread sticks to a convenient spot. It then strengthens this primary thread with extra strands. When it is sufficiently robust to take the load of the whole orb, the spider hangs a second line in the form of a Y below the primary thread, making up the first three radials of the orb. Once all the radials are in place, it makes a temporary spiral before constructing the final, more finely pitched spiral, eating the temporary spiral on the way.

Web visibility

The human eye is incapable of detecting objects with a diameter smaller than 25 microns at a distance of 4 inches (10 cm), but the average diameter of a thread of orb web is around 0.15 microns, the thinnest threads being only 0.02 microns thick. The only time we can see the thread is when it is covered with dust or dew or when sunlight or another bright light source catches it. This is why much of the orb web's construction is invisible to us; only portions of the web catch the light at any one time. For this reason diagrams show the construction better than photographs.

▼ Construction of a typical orb weaver web.

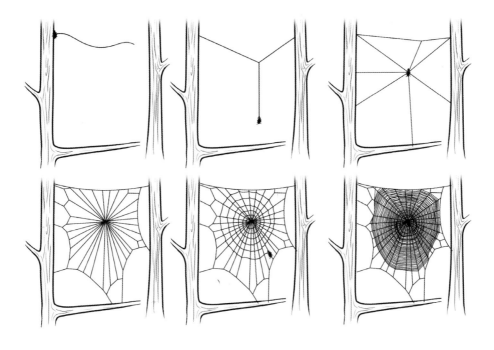

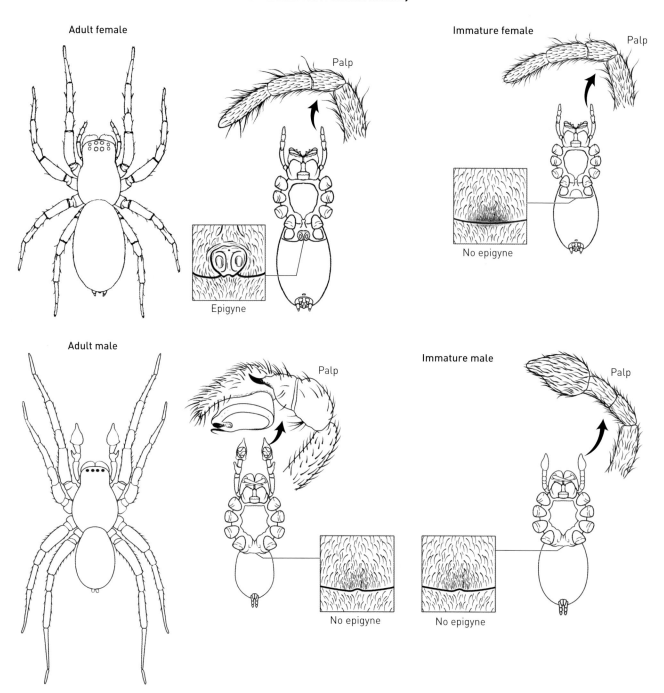

Adult female

Palp

Epigyne

Immature female

Palp

No epigyne

Adult male

Palp

Immature male

Palp

No epigyne

No epigyne

Sexing a spider

Spider identification is not a priority here, as several books accomplish this extremely well (see bibliography), but it is still very useful to be able to determine the sex of a spider, especially as this is the first step in their identification. Many species of spiders, especially the smaller ones, can be identified with certainty only by careful examination of the palps or epigyne under the microscope. Immature specimens are best tossed away!

Courtship and mating

The courtship and mating practices of spiders are among the strangest acts in the animal kingdom. The genital organs, their physiology, the elaborate courtship and the bizarre copulation behavior that follow would seem to stem from science fiction rather than planet Earth. Male and female spiders are often strikingly different in color, size or shape; commonly adult males are often smaller than females and can be readily recognized by their conspicuously large palps.

The genital openings of both sexes are located on the underside of the abdomen between the book lungs. In addition, the male has extraordinarily complicated ancillary sex organs on the terminal joint (tarsus) of each palp. The anatomy of these structures is completely different for each species, and there is no connection whatsoever between these and the testes in the abdomen. The genital organ of the female, the epigyne, is positioned just above the genital opening (epigastric furrow) between the book lungs. Its complexity does not compare with the male's equipment, but the important point is that the two are designed to interlink perfectly during copulation, rather like a lock and key.

The initial stages of the long drawn-out mating process work as follows. The male builds a sperm web, a small rectangle or triangle of silk onto which he deposits a small drop of seminal fluid containing the spermatozoa. Now the palps come into play as they dip into and suck up the liquid rather like a fountain pen filling with ink—whether this operates by suction or capillary action is uncertain. The sperm is stored there until the male finds a mate.

The male's next task is to look for a mate. This may take anything from a few minutes to several days, depending on the type of spider and the density of the local population, but once the male senses the presence of a female, the game is on. In some spiders the courtship may be very brief or nonexistent—they simply fling their legs over each other and mate—but in the majority of species it has evolved into an elaborate ritual often surpassing the most inventive and flamboyant displays of birds.

There are sound reasons for this ritualistic behavior. Mating is often a particularly dangerous time for male spiders, as not only are the female's instincts strongly predatory but she is also usually considerably larger than her mate. Unless the male approaches her in the right way and adopts the prescribed prenuptial ritual, he

Epigyne of female spider (*Pisaura*).

A male orb weaver (*Neoscona*) tentatively approaches the female.

stands a grave chance of being mistaken for prey and ending up as a meal instead of a mate.

The nature of courtship depends on the lifestyle of the spider and the importance attached to its various senses. In the case of web builders, the web acts as the communication line during courtship. The male begins by delivering a series of gentle tweaks to the web in such a manner that the female recognizes the coded signals. If she is happy, the male is allowed into her home and mating can take place. In the unlikely event that the incorrect code is transmitted or she is not ready, the male will have to watch his step. One genus of tropical salticid spiders, *Portia,* has learned to take advantage of this by mimicking the coded signals of male web-spinning spiders and so gaining access to the female, which is promptly attacked and eaten. *Portia* even seems to have a genetically programmed built-in database of signals from which it can make a selection according to species. It is also capable of flexible trial-and-error adjustment of signals in response to feedback from the prey. Whereas the preprogrammed repertoire of signals is consistent with animals governed by instinct, trial and error is an example of problem-solving behavior that is less expected in an invertebrate.

Hunting spiders depend on sharp eyesight for finding prey, so they have evolved a visually based courtship. Once the male has found a suitable mate, perhaps by tracking her down by pheromones in a similar way to some insects, he signals his intentions by a complicated display of leg and palp movements. The spiders that win all the prizes for chivalry are the enchanting and perky little jumping spiders. As

➤ The male inserts his palps into the female's epigyne. Note the inflated palps.

they possess the most acute eyesight of all, they perform the most elaborate dances, involving much waving of legs and vibration of palps, the nature of the display being specific to each species. Such spectacular performances may be enhanced by adornments of bright metallic colors and ornamental tufts of hair on head and legs.

Other families adopt simpler courtship strategies. For example, among the relatively primitive mygalomorphs, where the female lacks an epigyne and the male has simple palps, courtship is generally done by touch alone. Then there are spiders that may emerge only at night and live underground under stones or bark or in caves, which gain information about their surroundings almost entirely by touch or taste, using special sense organs on their legs. It is unlikely that such encounters are left to chance, as this would lead to an evolutionary blind alley, especially when populations are at a low point for some reason. Such spiders, like the hunting spiders, probably rely on pheromones.

Some male spiders, such as the nursery-web spiders *Pisaura,* practice a form of bribery by presenting their mate with gifts of wrapped prey, so distracting their partner's predatory instincts elsewhere. One of the most freakish techniques is adopted by *Xysticus* crab spiders, which employ what can only be described as bondage. Here the male circles around and over the female, stroking her gently with his legs until she is in a submissive state. As he does so he covers her with a thin veil of silk, effectively immobilizing her so that he can mate without risk to himself. After the male has left she escapes from her bonds to lay her eggs.

As with courtship, the physical act of copulation varies widely among different families. The main object of the exercise is for the pair to adopt such a position that the previously charged male palp is coupled with the female's epigyne, so allowing the seminal fluid to be transferred. The details of the procedure depend largely on the complexity of the palps and the epigyne, but it is beyond the scope of this book to describe the physiology of these here. The spermatozoa are stored in the female's abdomen until needed to fertilize the eggs.

Some male spiders ensure their paternity by fabricating a biological chastity belt by sealing their mate's epigyne. In other species the same object is achieved by the couple simply remaining together in harmony, leaving the male to repel other would-be suitors. The males of many spiders leave their mates to wander about looking for other females to couple with, but as they become weaker they may fall prey to another female of the same species. As humans we may find this horrific, but from the spiders' perspective such a fate is far better for the survival of the species than being snapped up by a bird or simply dying of exhaustion or old age.

Spider bites

Unlike those of the majority of insects, the bodies of spiders are soft and easily injured. Also, spiders are not equipped with claws, large mandibles or stings and most are weak and timid. The capacity to stop enemies and prey is critically important to them—hence the web and a venomous bite are crucial to spiders' survival. Paradoxically the most venomous species are typically quite small; they depend on their powerful venom to instantly immobilize dangerous prey.

Male water spider (*Argyroneta*) showing its large jaws.

Spiders rarely bite, even if provoked. To a spider human skin is merely another surface to walk over and so there is no point in biting it, besides which the vast majority of northern European spiders are far too small to be capable of breaking its surface. Their biting instincts are geared to respond to small moving or vibrating objects such as a fly rather than a finger. Most bites are the result of a spider becoming accidentally trapped against the skin.

Only a handful of European species have the potential of biting a human, and the effects are unlikely to be much worse than a small pinprick and perhaps minor irritation. In spite of the abundance of spiders, bites are extremely rare. We all know that in hotter parts of the world there are a few nasties around—the black widow (*Latrodectus*) and brown recluse (*Loxosceles*)

are notorious examples from North America, and another is the Sydney tunnel-web spider (*Atrax*) from Australia, which is reputed to be the most dangerous spider in the world. But there have been few confirmed fatalities as a result of spider bites, while bees and wasps kill thousands of people a year. Fortunately none of the potentially dangerous spiders are found in northern Europe.

The fallacies surrounding tarantulas start with their name, which arose during the Middle Ages from the village of Taranto in Italy, where a bite from a certain spider was blamed for a range of symptoms from severe pain and vomiting to spasms and exhibitionism. The supposed cure was to perform a frenzied dance—the tarantella—until the victim dropped from exhaustion. Curiously, the species of large wolf

Woodlouse spider (*Dysdera*) in threat pose.

spider allegedly responsible spends most of its life in an underground burrow; it's unlikely to bite anybody and its venom is relatively harmless to humans. The culprit, if there was one at all, may have been a theriidid, *Latrodectus tredecimguttatus,* an attractive red-spotted spider closely related to the black widow.

Even the large, hairy bird-eating spiders or tarantulas are misrepresented. Although their impressive fangs can easily break the skin, the venom from most species has little effect on humans. Perhaps one of the most famous tarantulas of all time was Thomas, a spider I owned in the 1960s, which was found in a bunch of bananas. In his starring role in the first James Bond film, *Doctor No,* Thomas was persuaded—albeit reluctantly, I should add—to walk over the naked chest of Sean Connery ... but that's another story!

That being said, there are a few exceptions in both northern Europe and North America. Some of the larger species such as the house spider (*Tegenaria*) and the garden spider (*Araneus*) are capable of giving a little nip if carelessly handled or imprisoned by a hand. The water spider (*Argyroneta*) and woodlouse spider (*Dysdera*), both large-jawed species, have a reputation for biting without much provocation, and there are a number of reports of mouse spider bites, but the reason is quite understandable in the latter case. The mouse spider is not only common but has a predilection for creeping around the walls of houses at night in search of prey. Come twilight it often takes refuge in garments scattered on the floor, especially in corners or against the floor edges. When the spider's slumbers are disturbed by being thrown against an armpit or some other sensitive part of the human anatomy, it naturally tries to defend itself by using the only weapon it possesses. I have been bitten by the uncommon *Araneus marmoreus* (page 102) in exactly these circumstances; the spider must have crept into my shirt while it was hanging out to dry on the clothesline near some bushes. It was only a pinprick, but if you are worried, shake out your clothes before dressing.

Another European spider capable of biting is *Segestria,* a huge creature with flashing green

A false black widow spider (*Steotoda nobilis*) lurking in the background of its scaffold web.

jaws that bites fiercely at anything hovering near the entrance of its characteristic tunnel in a wall—a finger, for instance (page 168). A friend of mine was brave enough to try this, whereupon the spider shot out like a moray eel, grabbed his finger and held on for more than 10 seconds before finally letting go. His finger remained numb for two days!

More recently there have been reports of an alien spider establishing itself in the United Kingdom, imported from Madeira and the Canary Islands; like the black widow spider it lives around houses and outbuildings. Its bite causes intense local pain and swelling. It is a theridiid spider a little smaller than the garden spider, but with a round, shiny dark brown body. Its name is the false black widow (*Steotoda nobilis*) (see above and page 148). Apparently its neurotoxic venom initiates production of a neurotransmitter, and it appears to mimic the venom produced by the black widow (*Lactrodectus*), to which this spider is closely related. It has been found in sheds and porches hanging upside down in tangles of web. The spider is clearly spreading—I found one on a bridge over a country stream some way from human habitation. Perhaps best avoided!

There can be little doubt that, apart from the bites of a very few particularly poisonous spiders, the physical effects of stings from wasps and bees are more painful and potentially more serious. For most people the psychological effects of a spider bite are more traumatic than the actual bite.

2
Nocturnal Hunters

The prey-catching techniques of nocturnal spiders, which hunt at night, do not depend on web traps or keen eyesight but rely instead on scent, touch or vibration to locate prey. Many spiders are particularly active at night, including many orb weavers, spitting spiders and house spiders, but the spiders described here are those that actively hunt at night. They are generally much duller in color than the daylight species, most being brown, black or gray, and are often furnished with fine hairs. Their habits too are less spectacular, as they prefer to creep about rather than making the high-speed dashes of daylight hunters. Their eyes are correspondingly much smaller than those of their more active diurnal counterparts. During the day these nocturnal hunters hide away in silken retreats under stones, in holes in logs and trees or curled up in leaves.

A common European nocturnal hunter—a species that is also spreading in North America—is the mouse spider, *Scotophaeus blackwalli*. It wanders about in houses at night looking for prey, then makes a short sprint and pounces on its victim. Another, less common species is the more brightly colored woodlouse spider, *Dysdera crocata*.

Nocturnal hunting families

The **Gnaphosidae** family, commonly known as ground or stealthy spiders, are mostly gray or black, lacking in pattern, and are furnished with short silken hairs, but some species are subtly iridescent. They have protruding and widely spaced cylindrical spinnerets.

Gnaphosids are primarily ground dwellers, only rarely occupying arboreal habitats. They spin tubular silken retreats in rolled leaves or under stones, hiding during the day and emerging only at night. These spiders rely on scent, touch and stealth to find prey.

Gnaphosids are often the most abundant spiders to be found in open and drier areas. About 250 species occur in North America and 15 in northern Europe.

◄ Woodlouse spider.

Sac spider (*Clubionia*) in cell.

Another family of nocturnal hunters are the **Dysderidae**, the woodlouse or long-fanged six-eyed spiders. Dysderidae are relatively primitive nocturnal short-sighted spiders, all of which have a rather long, smooth abdomen with no clear pattern or markings. In common with other ground spiders, they do not construct webs for catching prey but make silken retreats under logs and stones.

Their jutting jaws give them a menacing appearance. They possess six eyes only, which are arranged in a circular pattern. The female has no epigyne, while the male's palps are of simple design. This spider is a specialist in catching creatures that most other spiders reject or are unable to tackle—woodlice. Hence the impressive fangs, which are adapted to pierce the tough body armor of these animals and similar arthropods.

Four species occur in northern Europe, while only one is recorded from North America.

The **Clubionidae** family, sac or foliage spiders, are a largish family of mostly nocturnal spiders that bear a superficial resemblance to Gnaphosidae, the night prowlers. The easiest way to tell the two apart is to examine the spinnerets, which in Gnaphosidae are cylindrical and more widely separated; those of the clubionids are more pointed and generally appear smaller. In addition, sac spiders are typically foliage hunters. The majority of these spiders are brownish or gray, with little in the way of markings except for the subtle shadings of their velvety coat, although there are a few exceptions. Most species can be reliably identified only by microscopic examination.

Sac spiders are so named for their habit of resting during the day in a silken cell hidden in a rolled leaf among vegetation or under stones or bark. Some occur in dry situations at ground level, similar to the gnaphosids, while others prefer damper situations higher up in bushes and trees. North America has 58 species and 35 may be found in northern Europe.

Prowling or long-legged sac spiders, the **Miturgidae** family, are similar to the sac spiders and were once bundled together in the same family, but the former have longer legs and more robust bodies. Like sac spiders they are nocturnal wandering hunters that hide during the day in silken sacs. Most species are ground dwelling, living in forest, scrub and rocky deserts. This is a small family with only about 40 species worldwide. Twelve are found in North America.

The nocturnal spiders of the **Anyphaenidae** family, known as buzzing or phantom spiders, occur mostly in the foliage of trees and in leaf litter, hiding during the day in tubular silk retreats. Under a lens the family can be identified by the displaced tracheal spiracles, which are midway between the spinnerets and the epigastric furrow. Rather than trapping prey in webs, buzzing spiders hunt down insects in a similar manner to the running crab spiders, Philodromidae. There are 37 species in North America and one from northern Europe.

Mouse spider on electrical socket.

Mouse spider (*Scotophaeus blackwalli*)

Scotophaeus from Greek *scotos* (darkness); *blackwalli* after John Blackwall, a 19th-century spider expert

One of the most frequently seen spiders in houses is the mouse spider, a member of the Gnaphosidae family. Rather than spending its time hiding away on its web in dark corners like the house spider, the mouse spider stalks stealthily around the walls and ceiling at night in search of prey. It can often be seen in the form of a gently moving dark blob that pauses every now and again for a rest.

Although it does not build a web for trapping prey, like so many spiders the mouse spider trails a thread of silk as it stalks about, pouncing with speed and ferocity on any prey it encounters, such as mosquitoes, moths and flies. When not creeping around the house, the mouse spider spends the day in a silken cell hidden away in a crevice behind a picture or in the folds of curtains or clothing. Able to survive for months without water, it is well suited to a domestic environment.

A fully mature female mouse spider is about 0.5 inch (12 mm) long, and this size, coupled with her dark, velvety-sleek mousy appearance, gives her a slightly creepy aura. Although not aggressive, this spider has occasionally been known to nip more delicate areas of the skin. But such bites are insignificant when compared with the assaults of most other small creatures such as wasps and horseflies!

The mouse spider can be seen at any time during the year in the warmer parts of Europe, under bark or in holes in walls, but in England it lives only in and around houses. The species is now well established in North America, where it was imported from Europe.

Mouse spider peeping out from retreat in electrical socket.

A ground-living nocturnal hunter (*Drassodes*) showing typically somber colors associated with its lifestyle.

Ground spider (*Drassodes lapidosus*)

Drassodes from Greek "active on the road"; *lapidosus* from Latin "stony"

Another common but larger ground spider is *Drassodes lapidosus*. It is the largest and fiercest of the northern European gnaphosids, reaching about 0.7 inch (18 mm) in length. It is a sleek and mousy-looking spider with a pinkish gray abdomen and is lithe in movement, capable of moving rapidly when disturbed from its daytime retreat under a stone or log or at the base of a grass tussock. At night *Drassodes* emerges from its silken cell to prowl around like a panther in search of food.

There are several species of *Drassodes* in the genus, *D. lapidosus* being found all over Europe. Identification of some of them is possible only by careful examination with a powerful lens or microscope. There are six species of *Drassodes* in North America; one of the common ones is *D. neglectus*, a yellowish or light gray spider that has an indistinct pattern of chevrons toward the rear of the abdomen.

Fangs of woodlouse spider.

Woodlouse spider (*Dysdera crocata*)

Dysdera from Greek "without a fleece"; *crocata* from Latin "saffron yellow" (referring to its hairless yellow abdomen)

The bright orange legs and a carapace with contrasting cream abdomen, together with the massive protruding chelicerae with their wickedly sharp fangs, imbue the woodlouse spider, *Dysdera crocata*, with a sinister appearance. This member of the Dysderidae family (known as the sow bug killer in North America) favors slightly warm and damp habitats, often around buildings, spending the daylight hours in a silken cell under logs or stones. Come nightfall it wakes up to wander about in pursuit of suitable prey.

There are two similar European members of the genus, *D. crocata* and *D. erythrina*, and they are difficult to tell apart. The former species is also found sporadically in North America, where it was probably imported from the Mediterranean.

Most nocturnal spiders spend the day in a silken cell.

Sac spider on a nocturnal prowl.

Sac spider with prey.

Clubiona phragmitis

Clubiona from Greek "cage" (referring to the silken sac); *phragmitis* from Greek "reed"

The habits of spiders in the Clubionidae family are broadly similar to one another, although they occupy a range of slightly variable habitats, depending on species. Two European species are illustrated here: *Clubiona lutescens*, a common species that lives among low vegetation, shrubs and bushes, often in damp places, and *C. phragmitis*, which prefers drier habitats—this one was photographed in marram grass growing on a coastal sand dune. Neither of these two species occurs in North America, but there are many with similar appearance and habits.

Grasshead prowling spider with prey.

Grasshead prowling spider

(*Cheiracanthium erraticum*)

Cheirocanthium from Greek "thorned hand";
erraticum from Latin "wandering"

The spider illustrated, *Cheiracanthium erraticum*, is the commonest and prettiest member of this genus. Unlike other spiders of the Miturgidae family, which prefer a more arboreal lifestyle, it lives among low plants such as heathers and grasses. It also differs from the clubionids in having a longer, slimmer first pair of legs and narrowing, forward-slanting chelicerae.

One way to find *C. erraticum* is to look out for its silken cells interwoven with dead grass heads, which protect both the spider and her eggs; when backlit the spider can often be seen sandwiched within.

C. erraticum is widespread in Europe, while a similar species, *C. mildei*, occurs in both southern Europe and North America, where it was introduced. A few of the larger species have a reputation for biting people and causing necrotic blisters.

Buzzing spider showing characteristic double chevron marks on abdomen; with prey on flower (right).

European buzzing spider

(*Anyphaena accentuata*)

Anyphaena from Greek "without web";
accentuata from Latin "sings to others"

Few spiders make audible sounds, but the buzzing spider is an exception. As part of an elaborate courtship display, the male raises his front legs and violently taps his abdomen against the leaf beneath him, producing a high-pitched buzzing like a tuning fork on paper.

Anyphaena accentuata has the distinction of being the only member of the Anyphaenidae family found in northern Europe, but in North America, where the family has its headquarters, its relatives are quite common. *A. accentuata* can generally be recognized by distinct pairs of dark marks on the abdomen, although the female turns gray and loses her markings once she lays eggs.

Buzzing spiders are particularly adept at leaping rapidly from leaf to leaf, pouncing on insects such as small flies and leafhoppers that take their fancy. To help them cling to the leaves and twigs of shrubs and trees, they are equipped with special tufts of hair on their claws known as scopulae. The best way of finding a buzzing spider is to beat the lower branches of trees and shrubs, particularly oak trees, in early summer, although few males survive into June.

3 Daylight Visual Hunters

When compared with most other groups of spiders, the prey-catching methods used by the daylight hunters are relatively conventional. Rather than relying on cunning techniques such as web traps or spitting, they locate prey visually and run them down or ambush them. Many wolf spiders, for instance, execute this at high speed, rather like cheetahs chasing antelope, with the odd little jump thrown in from time to time. These spiders, which include mainly the jumping, wolf, nursery-web and lynx spiders, rely heavily on their eyesight for prey capture and so can generally be recognized by their large eyes.

The wolf spiders are the most familiar—anyone who has wandered around the countryside in spring or early summer will not have failed to notice restless movements on the dry, leafy woodland floor or open fields as these ground-loving hunters sprint off to safety. Indeed, any spider seen dashing around on terra firma is most likely to be a wolf spider. As most wolf spiders spend their lives on the ground, they tend to be brown or gray in color, although some species are blessed with brighter markings that become evident only when examined closely. Almost all wolf spiders hunt in daylight, so they need keen eyesight, this being provided by the two large, forward-pointing median eyes. Although frequently found in large numbers, they do not hunt in packs, as their name suggests!

Other spiders with a similar predatory lifestyle include the hunting spiders, Pisauridae—the best known of which is the handsome nursery-web spider, *Pisaura mirabilis*—and the lynx spiders, Oxyopidae. The latter hunt actively on vegetation, often leaping from leaf to leaf like jumping spiders. Three species of lynx spiders are found in Europe but only one member of the family lives in England, where it is rare and restricted to a few Surrey heathlands; more colorful and larger species are found in North America. Another family included here are the ghost spiders, Zoridae. These resemble sac spiders in some respects but unlike them are basically daylight hunters.

◄ Lycosid wolf spider showing eye layout (*Trochosa*).

Daylight visual hunting families

The dark, furry appearance of **Lycosidae**, or wolf spiders, together with their ability to chase down prey at high speed, has earned them their family name. They need excellent eyesight for their active hunting technique and so have especially large anterior eyes set in an arrangement that is characteristic of the family. It is one of the largest groups of spiders, with about 3,000 species worldwide, superseded in numbers only by jumping spiders (Salticidae), money spiders (Linyphiidae) and comb-footed spiders (Theridiidae).

Wolf spiders are largely free-living and can be seen scurrying over the ground as they run for safety at our approach, especially on warm days. The females can often be spotted carrying their egg-sac attached to the spinnerets. Some dig burrows and pounce on or give chase to insects that wander close to the entrance. Some wolf spiders—*Trochosa* species, for instance—are nocturnal and can be detected at night with a flashlight, as their eyes reflect the light back in the same way as those of nocturnal mammals.

Like wolf spiders, the **Pisauridae**, or nursery-web spiders, are free-living, so they do not build webs to trap prey but rely on their excellent eyesight instead. They are usually larger than most wolf spiders, varying from 0.32 to 1.2 inch (8–30 mm) long. Females build large, conspicuous tent-like webs for protecting their offspring. Rather than carrying her egg-sac on her spinnerets like the wolf spiders, the female uses her jaws. This family includes the raft or fishing spiders described in chapter 9.

Like the wolf and nursery-web spiders, lynx spiders, or **Oxyopidae**, are agile daytime hunters, but even more so, relying on their large and efficient eyes to find prey. Once the target is located, lynx spiders slowly creep forward, cat-like, until within pouncing distance—which can be quite long. Indeed, their jumping abilities can sometimes almost match those of the true jumping spiders. They may also be found running and leaping through foliage and flowers in active pursuit of prey, sometimes stopping to crouch low before continuing their quest. They are largely a tropical family.

Lynx spiders are sun lovers and are often beautifully camouflaged. Most can be identified by their heavily spined legs, rather slim, pointed abdomen, high-domed head and eye arrangement, with one smaller pair positioned below a hexagonal pattern of six larger ones. North America can boast of 18 species, but Europe has only three.

To quote *Spiders of North America* by Ubick, Paquin, Cushing and Roth, "zorids have wandered about (in taxonomic terms) since the mid-1800s and have yet to find a comfortable placement." In the past **Zoridae**, the ghost or spiny-legged spiders, have been classed at various times with the Gnathosidae, Clubonidae, Lycosidae and Ctedinae. For the time being they have been given a family of their own, but maybe DNA analysis will reveal their true family origins.

Zorids are sprightly ground- and shrub-dwelling spiders that are active mainly during the day, having a similar hunting strategy to the clubionids. Ghost spiders do not build webs but hunt actively, chasing down their prey among low vegetation or at ground level in leaf litter, moss and the detritus of hedges and woods in a variety of dampish habitats. Seven species occur in northern Europe but only one in North America.

➤ Nursery-web spider (*Pisaura mirabilis*) guarding nest of spiderlings.

Spotted wolf spider carrying spiderlings on its back.

Spotted wolf spider (*Pardosa amentata*)

Pardosa from Greek "spotted like a leopard";
amentata from Latin "furnished with a strap"

The majority of Lycosidae wolf spiders carry their egg-sac attached to the spinnerets, which makes the females very conspicuous as they run about on the ground. This exposes the eggs to the warmth of the sun, accelerating their development. On hatching, the spiderlings climb onto their mother's back and are carried around by her for about a week. When viewed with the naked eye the female's abdomen looks fuzzy and irregularly shaped; a close inspection will soon reveal the individual spiderlings. The males are a little smaller than their mates; like many wolf spiders during courtship, they employ a system of semaphore, signaling with the palps and front legs.

Pardosa amentata is a ubiquitous species, being found in a wide variety of habitats—this one was living among a patch of wildflowers in my garden. It is one of the commonest species in Britain and can be seen any time from spring to autumn in northern Europe.

Common pirate wolf spider on pond surface with egg-sac.

Wolf spider (*Pirata piraticus*)

Pirata from Greek "sea-robber"; *piraticus* from Latin "piratical"

This velvety rust-brown wolf spider, with its white spots and smart white lateral stripes running down the full length of the body, is a lover of watery habitats. In hot weather this member of the Lycosidae family can often be seen running over the surface of ponds and bogs, where it hunts for insects on or just below the surface.

The female here is basking in the midday sun with her egg-sac attached to her spinnerets. When alarmed she will vanish beneath the surface, not to reappear for several minutes.

In North America this spider is known as the common pirate. Unfortunately this causes confusion with pirate spiders from an entirely different family, Mimetidae—spiders that really do lead a piratical lifestyle. Several very similar *Pirata* species are found in both Europe and North America.

Sand-dune wolf spider with egg-sac.

Sand-dune wolf spider (*Arctosa perita*)

Arctosa from Greek arktos, "bear"; *perita* from Latin "skillful" or "experienced" (at camouflage?)

Living on sand requires special adaptations and camouflage, and this lycosid species clearly exhibits both. Attractively decorated with subtle pink and black markings and annulated legs, *A. perita* blends perfectly into its natural habitat of coastal dunes and light sandy soil. In North America the very similar *A. littoralis* is sometimes known as the sand-runner.

Perita spends most of its time inside a silk-lined burrow that it excavates in the sand. When the weather is sufficiently warm it peeps out of the entrance, waiting to pounce on some unsuspecting insect that may wander by. Members of this genus have a rather flattened carapace with the posterior eyes mounted on top so they look upward. Like all wolf spiders, *perita* has acute eyesight that can spot the slightest movement. At any sign of danger, for

◄   Sand-dune wolf spider leaving its burrow.

instance, a human walking past several yards away, it will bolt down its hole, not to appear again until the potential threat has vanished. If the danger seems especially grave, this spider will seal the burrow entrance with a curtain made from sand and silk. Recording its activity on film entailed lying flat in the sand with a long-focus macro lens and keeping stock-still for an hour or two. Just a twitch of a finger sent the spider bolting back down the hole.

Unfortunately for *perita,* it has a deadly enemy, one that is capable of detecting it even when buried out of sight beneath the sand in its closed burrow. This is a small hunting wasp, *Pompillus plumeus,* that depends on the spider to lay her egg on, thus providing fresh meat for the larva that hatches. When the wasp senses a spider underground, presumably by scent, it digs frantically down into the tunnel to find the spider—"like an excited terrier," according to W.S. Bristowe. *Perita* has a trick up its sleeve, though, which more often than not enables it to make an escape. The tunnel is Y-shaped, allowing the spider to execute a high-speed sprint through the alternative fork into the open—followed no doubt a second or so later by a frustrated wasp.

Wolf spider (*Arctosa cinerea*) with prey at entrance of its riverbank burrow.

Wolf spider (*Arctosa cinerea*)

Arctosa from Greek *arktos*, "bear"; *cinerea* from Latin "ash-colored"

This splendid lycosid is Britain's largest wolf spider, although both sexes lack the brighter colors of some of their smaller relatives. The female is shown here. It spends most of its life in a silk-lined burrow of its own making in the sand and pebbles of rivers and lakesides, where it waits in ambush for any invertebrate that comes within its field of vision. From time to time the spider will also venture out of its hiding place to hunt more actively.

Most *Arctosa* species are mottled with gray or brown markings and have annulated legs so that they blend into their preferred habitat among sand and stones. *Arctosa cinerea* does not occur in North America, although there are a number of similar species. One way of spotting these spiders is at night. If you hold a flashlight level with your eyes, the spider's eyes will reflect the light with a characteristic blue-green glow.

➤ Wolf spider basking on riverbank shingle.

Female nursery-web spider with egg-sac.

Nursery-web spider (*Pisaura mirabilis*)

Pisaura from Latin *Pisaurum* (Pesaro) in Umbria, Italy; *mirabilis* from Latin "wonderful" or "extraordinary"

This large, handsome spider is a familiar sight to country dwellers all over England and northern Europe, where it can be found around heathland, grassland and woodland clearings. A very similar species, *Pisaurina mira*, is found in North America.

The most obvious evidence of *Pisaura*'s activities appears during late summer, when conspicuous nursery tents can be seen strewn among the low vegetation of their favorite haunts.

These serve to protect the egg-sac, which the female hangs up inside while she stands guard outside, ready to attack any intruder that dares to get too close.

This spider's color ranges from rich chocolate brown or light tan to light gray, with a pale narrow band running down the middle of the carapace and an abdomen bordered by wavy lines. There are three northern European species of hunting spiders in the Pisauridae family, all of which have long, robust legs and acute eyesight. All are active hunters either in low vegetation, like *Pisaura* and *Pisaurina,* or on the surface of water, as in the case of *Dolomedes* (described in chapter 9).

Like other hunters, the female does not spin webs to catch prey but runs around among low plants or on the ground in pursuit of her quarry. When at rest or sensing prey, she frequently can be seen with her two front pairs of legs held together and extended stiffly forward at an angle, like a pointer sniffing the air. Unless you approach very cautiously she will dart off to hide under a leaf or jump down into lower vegetation. During courtship the male presents his mate with a wedding present of a wrapped juicy grasshopper or some other insect, which acts as a diversion during mating. However, he has been known to cheat by wrapping up an empty carcass or even running off with the gift at the end of the mating ceremony!

Come July, the female may be found trundling about with a large spherical egg-sac beneath her sternum. When the eggs are ready to hatch she attaches the sac to some low vegetation, such as long grass or heather, and weaves the large protective tent all around it. On hatching, the spiderlings cluster together for a few days before molting and gradually wandering off on their own.

Female nursery-web spider.

Male nursery-web spider.

A lynx spider from Northern Europe (*Oxyopes heterophthalmus*).

Lynx spider (*Oxyopes heterophthalmus*)

Oxyopes from Greek "sharp"; *heterophthalmus* from Greek *heteros ophthalmos*, "different (other) eye"

Most of the 500-odd species of lynx spiders, the Oxyopidae, live in the tropics, although a few are found in Europe and North America. Britain has only a single representative, *Oxyopes heterophthalmus*, which is unfortunately rare, being restricted to a few heathland areas in Surrey, although it is widespread over much of Europe. One particularly attractive species from the southern United States, the green lynx spider, *Peucetia viridians*, will spit venom at any intruder that threatens her or her progeny from up to 8 inches (20 cm) away.

The green lynx spider from North America.

Green lynx spider (*Peucetia viridians*)

Peucetia—alternate name for Pasithea, one of the three graces of Greek mythology; *viridians* from Latin *viridis*, "green"

This handsome, vivid green spider is common in the southern United States and Mexico. Larger than its European counterpart, it is only a medium-sized spider measuring about 0.5 inches (15 mm). Although those from the southeastern states are bright green, specimens from the western side tend to be yellow or brown.

Like the European species *Oxyopes heterophthalmus,* this spider hunts prey by day, running with agility and leaping from one stem to another. It can also adopt a more passive approach by waiting for insects on flowers and stems, sometimes standing on its hind legs with front legs raised in a posture reminiscent of a praying mantis.

The female spins a large egg sac and extends lines of silk to nearby vegetation, forming a sort of nursery web. Here she stands guard, ready to spit venom into the face of any intruder.

Close-up showing characteristic eye arrangement of ghost spiders.

Ghost spider (*Zora genus*)

Zora from Greek "violence"

Ghost spiders have a certain stylish charm of their own, although their full beauty can be appreciated only when examined under a magnifier; this reveals a background of pale yellow embellished with subtle darker brown striations.

The pointed carapace perhaps best identifies the Zoridae, together with their large, dark eyes set in two curved rows, the posterior row being so rounded that they almost appear to be in three rows. Most species are virtually impossible to tell apart without a microscope. Another characteristic of ghost spiders is their capacity to sprint at high speed and to jump when the need arises—abilities shared by many spiders, but perhaps with less elegance.

Ghost spiders normally live amid leaf litter and debris on ground level, although the one illustrated was found on a garden shrub.

4

Jumping Spiders

Strictly speaking, jumping spiders should be bundled with the "Daylight Visual Hunters," but as their attributes are so exceptional, together with the fact that the family is by far the largest, I have accorded these spiders a special section of their own in this book.

Jumping spiders, or Salticidae, are the most fascinating and advanced family of spiders, capable of charming even hardened arachnophobes. Indeed, these spiders have evolved such exceptional physiological and behavioral abilities that the family has become the largest in the spider world, with more than 5,000 species named so far. Apart from their athletic prowess, some of the tropical species are the most colorful and bizarre-looking spiders to be found anywhere. A glimpse of a metallic red and blue *Chrysilla* bouncing through a patch of sunlight like a sizzling spark in an Old World jungle is an unforgettable experience.

These alluring spiders are also associated with bright sunshine, in which they may be seen walking or skipping on a warm summer's day. Jumping spiders are unmistakable in appearance, having two large, owl-like forward-facing eyes protruding from a square-faced head, as well as six other, smaller eyes. Sharp eyesight dominates their sensory universe, as it does ours. There is something beguiling about the way they watch us, swiveling their heads from side to side or up and down, sometimes following our every move. Vision also dominates their courtship behavior, which involves much waving of their often brightly colored legs and palps.

Jumping spiders are hunters that walk around in a series of jerks while spotting for prey from a considerable distance. Once the intended quarry is detected, the spider stealthily approaches until it is sufficiently close to leap onto its victim's back. Spiders from some other families are capable of jumping short distances, but only salticids make accurate vision-guided leaps onto prey or other objects. Over a narrow angle their spatial acuity is said to exceed that of large dragonflies by tenfold.

◄ Fence spider (*Marpissa muscosa*) about to land. Note the safety line.

Unidentified jumping spider showing the impressive anterior eyes of salticids, and the bright palps, which are used for visual signaling.

It is this acute eyesight rather than their jumping prowess that makes these engaging little creatures so special. The optical and neurological equipment by which this is controlled is contained in the characteristic large, dome-shaped cephalothorax.

The eyes of salticids are physiologically unique among arthropods. As well as possessing high-resolution, full stereoscopic color vision, they are able to adjust the angle of view and focus by moving components inside the eye, including the retina itself! If you gaze into the eyes of a salticid you may be lucky enough to see a mysterious flicker or color change as the retina is moved. The eyes are constructed rather like binoculars, with a long- and a short-focus lens at each end of a long tube and a layered retina at the rear of the cephalothorax. This sophisticated mechanism allows the spider to locate, track, stalk and leap onto active prey. Using just visual cues, salticids can discriminate between

prey and predators, mates and rivals. Recent research even suggests that their brains could be far more advanced than those of other spiders. Among other "intelligent" skills they have an almost uncanny ability to figure out complex paths to gain the best vantage points before pouncing on prey, much like lions.

Some salticids can jump more than 20 times their own length, and these Olympian leaps are powered not directly by muscles but by hydraulic action. All the internal organs and legs are immersed in a blood-like fluid, so when a large muscle in the cephalothorax suddenly expands, the legs are rapidly extended through hydraulic pressure. One would hardly imagine that this could result in an accurate jump, but it usually does, as proven by high-speed photography.

Around 315 species of jumping spiders have been recorded in North America and 75 in Europe.

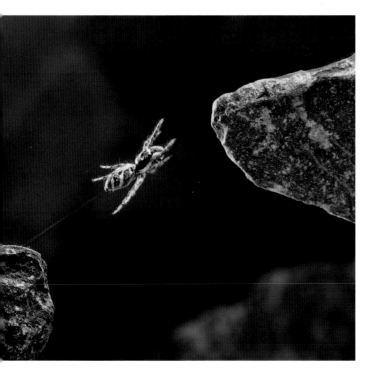 

Sequence showing zebra spider during the latter stages of jumping. Note the safety lines, including one on the left from the previous jump.

Zebra spider (*Salticus scenicus*)

Salticus from Latin *saltus* (leap); *scenicus* from Latin "actor"

This charming little black-and-white-striped spider is the best known of the large Salticidae family of jumping spiders. It is found not only over much of North America but also in Britain and northern Europe. The zebra spider can be found almost anywhere that is sufficiently warm and sunny, especially around houses and gardens on walls, fences, plant containers and windowsills. Less frequently it is found away from human habitation, on sun-exposed rocks and sunny tree trunks, moving over the surface in its typically jerky manner.

When the warmth of the sun disappears, the spider vanishes too and hides away in some suitable crevice. Evidence of its activities can often be seen on the surfaces it frequents, in the form of crisscross strands of web—like so many spiders, it leaves a trail of draglines wherever it goes.

As in all salticids, the eyes of the zebra spider are spectacularly acute. Although the smaller eyes may not have high resolving power, they do provide a 360-degree field of view to spot any movement. Once some visual disturbance is detected, the spider will raise its head and orient itself so that the main anterior median eyes (the headlight ones) scan the potential prey in detail. Its gaze will even latch onto minute prey such as a greenfly offered on a pair of fine forceps from several inches away.

The series of photographs of a female shows the action of the hydraulically powered leap in detail, including the dragline and its anchor point. The relaxation of the flexor muscles of the two pairs of hind legs that provide standing-start takeoff power can be seen in the picture on page 69. Adult males have huge, unwieldy chelicerae that are used for elaborate courtship displays as well as bloodless sparring contests between rival males.

There are four similar species of salticids in the family, of which two are rare in Britain.

Fence spider (*Marpissa muscosa*)

Marpissa from Greek *marpto*, "to seize";
muscosa from Latin "mossy"

Although not a common species, the fence spider is the largest salticid to be found in Britain, adult females reaching 0.4 inch (10 mm) in length. Its natural habitat is around stone walls and the trunks of trees, especially those that are exposed to warm sun and have peeling bark, under which these delightful spiders hide away in their whiter-than-white silken cells. Man-made wooden structures such as fences are just as popular, particularly those with peeling bark or clefts in which to take refuge and protect their cocoons.

Marpissa is so well camouflaged that it is almost impossible to spot when resting on bark. Perhaps the best way of finding this spider is to examine the south-facing surface of a gate or fence, where it is much easier to spot while basking in the heat of the sun. Remember, though, to approach very slowly, as the super-sensitive eyes will readily detect your presence and the spider will dash off to hide on the other side or in some crevice. Although the fence spider does not occur in North America, other, similar species in the genus are found there.

The series of photographs clearly shows the action of the jump. Note the trail of thread that acts as a safety line in case of misjudgment.

➤ From top to bottom, the jumping sequence of the fence spider.

▼ The eyes from the side. Note the flat head for this species' crevice-dwelling life.

This elongated Marpissa species lives amid grass stems.

Marpissa nivoyi

Marpissa from Greek *marpto,* "to sieze";
nivoyi from de Nivoy, a 19th-century spider researcher

Another *Marpissa* jumping spider is *M. nivoyi*. It is more elongate than the fence spider, with a rather antlike appearance, and is adapted to a life among grass stems. Like the fence spider, its front legs are thickened.

This spider is rare in England; it is most likely to be found among marram grass on coastal dunes and, less often, farther inland in marshy areas. When not stretched out along a blade of marram, it lies concealed in a silken cell within the hollow stem. The spider has the rather non-spider-like trait of sometimes running backward like a pseudoscorpion, another arachnid that is often found in the same habitat.

A similar species, *Marpissa pikii*, is found in North America and favors similar surroundings.

A sun-loving jumping spider (*Heliophanus cupreus*), caught a split second before jumping.

Heliophanus cupreus

Heliophanus from Greek *helio*, "sun"; *cupreus* from Latin *cuprum*, "copper"

Most of the *Heliophanus* genus of salticids can be readily identified in the field, as the females of most species have a striking color combination: a black body with pale yellowish-green legs. The males, though, are less impressive, tending to have darker legs and a subtly iridescent body—the species name *cupreus* refers to this spider's coppery appearance in bright sunlight. Both sexes have a thin white band around the front edge of the abdomen and white spots on each side of the dorsal surface.

As its generic name suggests, *Heliophanus cupreus* is a sun lover, like the majority of the family. It can be found on low vegetation and is most active near the top of plants in hot sunshine. This spider can be differentiated from other *Heliophanus* species by the black streaks down both sides of the femur and tibia on all four pairs of legs. Although absent from North America, the species is common and well distributed throughout Europe.

Male *Evarcha arcuata*.

Evarcha arcuata

Evarcha from Greek; *arcuata* from Latin
"curved" (referring to the shape of the spider's
epigyne)

This charming little jumping spider, along
with the rest of its genus, shows wide variation between the sexes, as is clear from the
photographs. Whereas the male E. *arcuata* is
dark brown or black, the female is a tawny
color marked with chevrons and thickly covered with white hairs.

◄ Female with earwig.

The courtship of spiders is often an elaborate performance, and that of jumping spiders
is especially so. *Evarcha* species are quite ener-
getic, with much waving of palps and front
legs. After mating, the female lays her eggs
within rolled-up leaves or, as is usually the case
with *arcuata,* in sprigs of heather tied together
with silk, and there she stands guard until the
eggs hatch. Maturity is reached in midsummer,
when the best place to find them is heathland
in southern England, where these salticids are
sometimes quite common.

This spider is widespread throughout
Europe but absent from North America, where
there are several other *Evarcha* species.

◄ ▲ Jumping sequence of *Phlegra* attempting to catch a mosquito.

Phlegra fasciata

Phlegra from a city in ancient Macedonia;
fasciata from Latin *fascia*, "band"

An inhabitant of coastal shingle and low vegetation around anthills, this dapper little dark-and-light-brown-striped jumping spider is difficult to spot unless seen moving. The male has a glossy abdomen and much more subdued markings than the female. The series of photographs shows a juvenile female leaping toward a mosquito (which actually managed to get away). It is clear that the spider shows no hesitation about tackling prey larger than herself—not that a mosquito poses much threat to her.

Phlegra is both rare and localized in Europe and North America. In England this salticid has been recorded only on a very few stretches of coastline in the south.

Face of male magnolia green jumping spider.

Magnolia green jumping spider
(*Lyssomanes viridis*)

Lyssomanes from Greek "raving mad" (referring to the spider's manic activity); *viridis* from Latin, "green"

This bizarre-looking, vivid, translucent green jumping spider from the southern United States hardly resembles a salticid at all and could be mistaken for a lynx spider at a casual glance. Both sexes have long legs and palps and are extremely active—even for jumping spiders. The awesome face of the male is dominated by his spectacular anterior median eyes. These appear to rotate in their sockets, but it is only the retinas, controlled by six pairs of muscles, that move behind the perfectly clear front lens. Adding to the inscrutable stare is their chameleon-like ability to move each eye independently.

Lyssomanes can be found living in all types of woodlands, particularly broad-leaved evergreens such as live oak and magnolia.

A pantropical jumper's eyes are alert to the slighest movement.

Pantropical jumper (*Plexippus paykuli*)

Plexippus from Greek "driving horses"

Many people who live in warmer regions of the world must be familiar with this natty and very active jumping spider. Not only has the pantropical jumper spread to most warm regions of the world, but it prefers to live around houses and other man-made structures, where it is readily noticed. Adding to its conspicuousness are its size—females are 0.4 to 0.5 inch (10–12 mm) long, large for a salticid—and the distinctive whitish bands that run down the entire length of the body.

The pantropical jumper was introduced to North America, where it is restricted to the warmer southern states. Unfortunately it is absent from northern Europe, although it survives well in the Mediterranean region. The species is very competitive, sometimes monopolizing structures such as old walls and even gas stations, where it tends to exclude all other species of jumping spiders.

◄ A pantropical jumping spider leaping inverted. Jumping spiders are perfectly capable of jumping from any position — this one has rotated 180 degrees prior to landing.

5 Ambushers and Lurkers

When considering the hunters and the ambushers, the overlap between the categories into which the spiders in this book have been divided is not always clear-cut. We have already seen that many of the so-called hunters frequently wait in ambush for prey, pouncing or giving chase at the last moment. By the same token, some spiders that are basically sit-and-wait predators can be divided into two groups. The first are those that sit in or close to their webs, lurking in nearby foliage or in tunnels that form part of the web; the web is instrumental in trapping or detecting the prey in the first instance. The second group does not rely on webs at all. Instead the spiders remain motionless, merging into the background until some unsuspecting creature falls into their arms, so to speak.

Crab or ambush spiders, the Thomisidae family, are masters at the second method. Instead of relying on webs, they use deception and camouflage, being adept at blending beautifully among the petals and stamens of flowers, where they wait motionless for their quarry. Some crab spiders are capable of changing color over several days to merge into the background yellow, pink, white or green of their chosen flower or foliage. Other species, especially those from the tropics, are able to mimic tree bark or even bird droppings.

Another group of spiders included here are the so-called running crab or small huntsman spiders, the Philodromidae, which are closely allied to the crab spiders (some experts bundle them into the same family with the Thomisidae). These too are capable of remarkable camouflage, remaining motionless and behaving much like the typical crab spider as they grab prey that ventures too close. However, as their name suggests, running crab spiders are usually much more active, being capable of running very rapidly in pursuit of prey or to escape predators. A common species of this type is the little running crab spider *Philodromus dispar* (page 89), which either chases prey or waits in ambush, employing both techniques with equal mastery.

◄   Male and female crab spiders (*Thomisius onustus*) prior to mating.

Female crab spider (*Misumena*).

In warmer parts of the world, including parts of North America, there are the larger and more scary huntsman or giant crab spiders from the Sparassidae family. In Europe this family has only one representative, a smaller and rather rare species, the green spider, *Micrommata virescens*.

Ambusher and lurker families

True crab spiders, the **Thomisidae**, have short, squat bodies with two pairs of atrophied-looking hind legs. By comparison the front two pairs are longer and more robust, designed to seize insects that wander too close. The males, in contrast to the sturdy-looking females, are slim, dwarf-like creatures. Crab spiders are lethargic by nature but, unlike spiders from other families, are capable of moving in any direction, like crabs. When alarmed they may be seen scuttling backward or sideways out of sight behind a flower or leaf.

Crab spiders come in a wide range of forms and colors, but most are crab-like, with an almost circular cephalothorax and a dumpy, squat abdomen; their two pairs of front legs turn inward and are considerably longer than the hind pair. They never build webs for trapping prey, preferring to lie motionless on a flower or leaf waiting for prey to visit, then grabbing it with the front pair of legs, which are held wide apart. Although thomisids have small chelicerae, the venom they produce seems to be highly toxic, as they have no trouble subduing large insects such as butterflies and bumblebees very rapidly.

There are 130 species in North America, but only 62 in northern Europe.

Whereas the Thomisidae are distinctly crab-like, the running crab or small huntsman spiders, **Philodromidae**, generally possess a longer, more oval abdomen; they have longer, thinner legs, with the hind pair nearly as long as the front pair, and are far more agile in their movements. In addition, to help these lively spiders clamber around actively in plants and execute lightning changes in position, they have scopulae on the soles of their feet. North America has about 96 species, while northern Europe has 62.

The **Sparassidae** family, or huntsman spiders (sometimes called banana spiders), are medium-sized to very large wandering spiders that are mostly tropical and rely on ambush to catch prey. Many species are crab-like and flattened, allowing them to creep into narrow crevices; they frequently get imported in banana consignments. Some species live in houses, where they are often encouraged, as they prey on household pests such as cockroaches. North America has 10 species, while northern Europe has only one.

Crab spider (*Misumena vatia*) awaiting prey.

Crab spider *(Misumena vatia)*

Misumena from Greek *miseo*, "hate or wrath";
vatia from Latin "bowlegged"

One of the most familiar crab members of the Thomisidae family—found in both North America and Europe—is *Misumena vatia,* known in North America as the goldenrod crab spider. Its color can vary from white to yellow, green or occasionally bluish, depending on what flower it chooses to adopt, and sometimes it exhibits light red spots or stripes. If *Misumena* moves from one sort of flower to another of a different color, it is capable of changing color to match the new surroundings; this spider is most frequently found on white or yellow blooms, oxeye daisies being a favorite. There it lies in wait for a visiting insect seeking nectar or pollen.

As an insect approaches, *Misumena* opens its two pairs of front legs and subtly aligns itself with the oncoming prey. Once sufficiently close the legs snap shut to embrace the victim, whereupon a bite is delivered without delay. Occasionally, if for some reason the toxic effects of the bite are slow to kick in, the spider will take flight on the back of a victim such as a large butterfly or bumblebee. However, such trips are short-lived—both spider and prey will tumble to the ground as the venom takes hold.

One way of finding flower-loving crab spiders is to keep a look out for immobile butterflies or bees on flower heads. The chances are that the unlucky insect is being held there by a crab spider.

These two pictures of the same crab spider species (*Xysticus cristatus*) reflect their choice of habitats.

Crab spider *(Xysticus cristatus)*

Xysticus from Greek "scraper"; *cristatus* from Latin "tufted" or "crested"

Many thomisid crab spiders are prone to variation, *Xysticus cristatus* being particularly so, as these two photographs show. This species lives among low vegetation or at ground level in a wide variety of situations from heathland to hedgerows. The exceptionally boldly marked specimen shown here was living on a Sussex sand dune, while the duller one with the fly was found among low-growing plants in an open forested area. This is the commonest and most widespread species in the genus and occurs throughout Britain and northern Europe.

Marsh crab spider, with front legs extended, climbing a plant.

Marsh crab spider (Xysticus ulmi)

Xysticus from Greek "scraper"; *ulmi* from Latin *ulmus*, "elm"

Similar to *Xysticus cristatus* is another crab spider, X. *ulmi*, but this one is less catholic in choice of habitat, preferring damp, marshy places among low vegetation. It too is distributed throughout Britain and northern Europe, and is replaced by similar species in the United States.

The most astonishing thing about *Xysticus* spiders is their courtship and mating behavior. Indeed, their technique is so bizarre that many experts doubted the validity of the behavior when it was first observed in the middle of the 20th century. The smaller and more long-legged male begins wooing his mate by putting her into a submissive frame of mind. He accomplishes this in typical spider fashion by gently stroking her with his legs. Once this stage is completed, he ties down her head and legs with bonds of silk to whatever she is resting on. He is now free to lift her abdomen, crawl underneath and mate—a process that may last up to 90 minutes. Once these matrimonial activities have been completed, the female breaks away from her bonds, which by now have served their purpose of preventing her from grabbing her suitor during the dangerous time shortly after mating.

Toad-like crab spider. Crab spiders often take prey larger than themselves.

Toad-like crab spider (*Oxyptila* species)

Oxyptila from Greek "sharp feather down" (possibly referring to the spider's clavate hairs)

Other common crab spiders in the Thomisidae family are those from the *Oxyptila* genus. These are similar to *Xysticus* but tend to have a more rounded and marbled abdominal pattern; like *Xysticus,* many show wide variation in colors and markings within the species. *Oxyptila* crab spiders are particularly squat and toad-like. Several species occur in both Europe and the United States.

During the day *Oxyptila* species are found at ground level, sometimes deep amid the base or roots of vegetation. They can easily be overlooked because of their effective brown camouflage and very slow movements. Also, like many spiders they are prone to feigning death by drawing up their legs and remaining motionless for several minutes. At night they may crawl up the vegetation, where they can be spotted with the aid of a flashlight. They are reported to feed during both day and night.

The green crab spider lives among leaves of trees and bushes.

Green crab spider (*Diaea dorsata*)

Diaea from Greek "during spring"; *dorsata* from Latin *dorsum*, "back"

The vivid green legs and cephalothorax and the yellow-margined, leaf-marked brown abdomen make this attractive medium-sized thomisid spider unmistakable. Like so many crab spiders, *Diaea dorsata* is beautifully camouflaged, making it extremely difficult to spot as it waits in ambush among the leaves of shrubs and bushes, especially young oaks. Individuals within a single species can vary widely in color and pattern, according to the color of the background they adopt. Some species are said to be able to change color gradually to match their selected leaf or flower. The spines on the front legs aid in trapping insects and are typical of crab spiders.

This spider is widespread in Europe but commoner in southern parts. In England it is rather locally distributed, mainly in the south.

Rear view of mating crab spiders. Note the enormous size difference between the sexes.

Crab spider (*Thomisus onustus*)

Thomisus from Greek *thomis*, "sting"; *onustus* from Latin "loaded" or "filled"

The most splendid of northern European crab spiders is *Thomisus onustus*. Adding to its allure is its comparative rarity—although it is widespread in Europe, the spider is found only in small numbers in England, being confined to certain heathlands in central southern parts of the country.

With her angular pink abdomen sporting two conical humps and her devil-like horned head, it is impossible to confuse the female of the species with anything else. The male is much darker and, in common with many other thomisids, much smaller, as is manifest in the mating sequence. Like *Misumena vatia*, this crab spider is able to adjust its color to match that of the flower upon which it rests.

T. onustus lives among mature heather, where it lies in wait. At the approach of an insect it will make smooth and barely perceptible adjustments to its position. As the potential prey lowers its head in search of nectar, it is seized in a flash. This spider often grabs insects considerably bulkier than itself, such as bees or even bumblebees.

6 Trappers: Orderly Webs

Whenever we think of spiders, the orb web springs to mind. It is the symbol of the spider, representing the pinnacle of its achievement in design and construction. The orb web evolved to create the maximum prey-capturing area while minimizing the amount of silk, but what is perhaps more remarkable is that the spider constructs this web in less than an hour. Only about 10 percent of spiders capture prey by means of orb webs or, in a few species, a section of an orb. Each web will exhibit subtle differences in design: there may be variations in size, orientation, number and distribution of radii or spirals, and ornamentation—the design of the stabilimentum, when present (see orb weavers). Such differences provide very useful clues to identification of the spider.

These intricate web designs are of course not learned but under strict genetic control—orb weaver spiderlings spin perfect orb webs soon after hatching, and the web remains fundamentally the same throughout the spider's life. The only differences that occur are those necessary to suit the size of the spider and the position of attachment points. Most webs are suspended vertically, but some are angled at 45 degrees or occasionally positioned strategically on a horizontal plane, depending on the type of prey they are intended to trap.

Orb webs often go unnoticed unless moisture condenses on their fine threads or they are backlit by the slanting rays of a low sun. Indeed, a dew-laden early autumn morning is the best time to enjoy their radial perfection. In these conditions the enormous number and diversity of spider webs that swathe almost every twig and leaf of hedge and meadow are quite breathtaking, and a reminder of the vast number of spiders that share our world—a very good reason for rising early on such days.

Once insects learned to fly—about 350 million years ago—it gave them a massive advantage in avoiding their enemies. Many spiders had no option, therefore, but to find a way of trapping prey in mid-flight. Thus, over the eons, spiders perfected the orb web, a device that captures flying insects with super efficiency. It was

◄ Orb web on autumn morning.

Web being extruded from spinnerets with aid of the rear claws.

no doubt a far simpler solution than evolving wings themselves.

The web itself is made from one of the strongest substances known. Although only a few microns in diameter, spider silk can stretch to many times its length before finally snapping. In this way the web absorbs the impact of rapidly flying insects with minimum risk of penetration. A further refinement in the construction reduces "bounce" so that the prey is not catapulted out of the web again. The insect's impact is confined largely to six radial spokes, which provide most of the forces that bring the web back to its original position. Recently it has been discovered that these spokes contain microscopic coils of web within the sticky drops along their length. These act as springs that help to stabilize the web, limiting bounce when prey hits it. There are also micro-

aerodynamic forces at work, based on wind resistance, that help the web to return quickly to its original position.

Different species position webs according to their favorite prey. The web's height, its angle and the type of habitat in which it is deployed are all crucial in providing the spider with as much food as possible. For example, some webs are installed near water to catch insects such as mosquitoes or midges, others may be angled low down in open meadow to trap leaping grasshoppers or leafhoppers, and others still are designed to catch larger flying insects such as butterflies and moths, so they are strung between shrubs and trees.

Only three spider families create orb webs. Most common are the webs made by Araneidae such as the garden spider. The other two families of orb weavers are the Tetragnathidae, or long-jawed spiders, and the Uloboridae. It is worth noting that a few spiders make orderly webs that are not at all based on the orb configuration—*Episinus* is a notable example. This genus belongs to the Theridiidae family, the scaffold-web spiders, which characteristically construct a web with threads that go in all directions. *Episinus* makes a much simpler web in the form of an H, with the two lower threads attached to the ground and the whole arrangement being held together by the spider (see page 139).

Orderly web-weaver families

Most spiders that build orb webs belong to the **Araneidae** family. The structure of their webs is subject to huge variations depending on species. Most are large in comparison with other orb-web spiders, almost all have a closed hub, and most of the webs are spun vertically. Whereas diurnal spiders dismantle and eat their webs in the morning, remaking them for daytime use, nocturnal species remake their webs in the evening.

Many orb weavers have colorful bodies marked with patterns of greens, yellows and reds, while others are drab brown. Most have rather short, spiny legs, which can help to distinguish araneid spiders from other families. As eyesight plays only a minor role in their daily life, araneid eyes are small. Their sense of touch provides most of the information they need about the outside world, particularly through contact with the web. Body size varies from 0.06 to 1.2 inch (1.5–30 mm). The Araneidae are a large family; 160 of them are found in the United States and 50 in northern Europe.

The **Tetragnathidae**, or long-jawed orb weavers, are closely related to the Araneidae, but most species have long bodies rather than round ones, sometimes up to two or three times longer than their width. Many species have enlarged chelicerae, particularly in the males.

Tetragnatha species constructs open-hub orb webs that are frequently oriented at an angle or in the horizontal plane. *Nephila*, on the other hand, makes a vertical bright yellow web, while *Meta* generally builds its web in dark places. Members of the *Pachygnatha* species construct webs only in the early stages of their lives, subsequently becoming vagrant hunters as they mature. When not in their webs, many species lie motionless, stretched out along a stem in a characteristic manner, where they are difficult to spot. This family has about 40 species in North America and 16 in northern Europe.

Uloboridae, the venom-free or cribellate orb-web spiders, are unique among spiders in two ways. First, they possess no poison glands; second, the various genera within the family construct entirely different webs. Some make complete orbs and others just triangular sections of an orb, while in some cases merely a few lines of silk are used. Also, unlike the Araneidae, rather than using sticky threads they make the capture area from fine cribellate silk.

The spiders themselves are generally smallish and have an unusual appearance, often with lumps or feathery tufts of hair. They can easily be mistaken for twigs or fragments of dead leaves. There are 240 species worldwide, 16 of which occur in North America and three in northern Europe. The generic name *Uloborus*, which comes from "lethal" in Greek, seems to be very inappropriate, as Uloboridae are the only family of spiders that do not possess venom!

The gold silk of the golden orb spider's web.

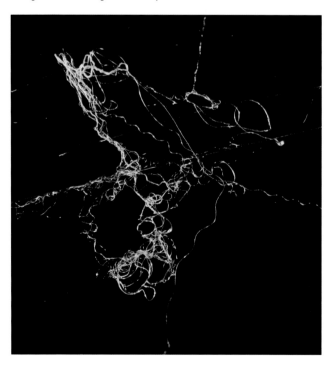

Hoverfly pictured a split second before flying into garden spider's web.

Garden spider; cross spider
(*Araneus diadematus*)

Araneus from Latin "spider"; *diadematus* from Latin "with a crown or diadem" (referring to the abdominal cross pattern)

The aptly named and ubiquitous garden or cross spider is the best known of all spiders, although it is not the most common spider to be found in gardens. During the Middle Ages the white cross on its abdomen boosted this member of the large Araneidae family into a creature of religious veneration, making it the paradigm of spiders. Sometimes it is extremely common in town and suburban gardens, among shrubs or on man-made objects such as window frames, where it is difficult to overlook because of its large size and tendency to sit bang in the center of its large orb web. Yet this species can be found in almost any habitat from heathland and woods to mountainsides and cliff faces. It is also found throughout the eastern United States, where it was introduced.

It is the female that is most frequently encountered, perched head down in the center of her orb. She is much larger than her mate, particularly in the autumn, when her spherical abdomen is hugely distended with up to about 900 eggs.

The orb web of the garden spider is the archetypal insect trap, designed for capturing insects as they fly around, oblivious to the invisible snare that awaits. The web, which is built on a near-vertical plane, is often large in relation to

In the autumn the garden spider is a common sight in urban gardens and towns.

the spider and has between 25 and 35 radii and close-set spirals. The center of the hub consists of meshed threads surrounded by a small spiral, beyond which is a free zone before the main spirals begin. It is these main spirals that have sticky droplets along their length for entrapping flying insects. The spider will either sit in the center of the hub with her eight outstretched legs each in contact with a radial thread, or hide away in a nook among some leaves, holding a stout line that is connected to the hub center. The exact position of a struggling insect will be detected at the slightest movement, whereupon she will be galvanized into activity, biting the victim before wrapping it up ready for her next meal.

The unrestrained ease with which spiders move around their webs without getting trapped themselves can be clearly observed by watching this large spider. This astonishing ability is accomplished in a number of ways. To begin with, she tends to sit facing outward with her body clear of the sticky spirals, so if the web is slightly inclined to the vertical she will settle on the underside. She will also move around by grasping the dry radial threads with her tiny claws, as only the spirals are sticky. Orb builders have special tarsi, with an extra third claw and opposing serrated hairs between which the silk thread is grasped. By twisting her feet at an angle to tension the thread, the spider will be held securely in position (see page 25). Additional immunity from entanglement is assured by an oily covering on her legs that further impedes adhesion.

Four colour variations of the four-spotted orb weaver.

Four-spotted orb weaver

(*Araneus quadratus*)

Araneus from Latin "spider"; *quadratus* from Latin "square"

The four-spotted orb weaver, along with the wasp spider (*Argiope bruennichi*), is the largest orb weaver found in northern Europe; an egg-laden female often attains as much as 0.6 inch (15 mm) in length. The male, though, like so many spiders, is much smaller. *Quadratus* is the only spider that might be confused with the garden spider, but when viewed from above, the circular abdomen with its square of four bold white spots should separate the two. It is also one of the most attractive species, with a color varying from rust red or dark brown to subtle pale shades of greenish yellow (see above).

A. *quadratus* is common and widespread throughout Europe. It lives in heather, tall grasses and low bushes such as gorse, where it builds a large orb web of about 16 inches

(40 cm) in diameter at heights between 3.3 and 5 feet (1–1.5 m) off the ground. Most of the day is spent hidden away in a sizeable retreat of plant material held together with tough, papery silk. Although this araneid is absent from the Western Hemisphere, North America has an almost identical species: the shamrock spider, *Araneus trifolium*.

To those who are tuned in to the grassroot jungle, a familiar sight in spring is the balls of little spiderlings of this species (as well as of the closely related garden spider and others). These remain intact for several days after hatching, making no attempt to move away or feed until disturbed by a jolt, a shadow or the warm breath of an enthusiastic observer, whereupon the little creatures scramble off in all directions. When danger has passed they gradually coalesce once again into a tight golden ball. Over a few days the ball expands until the spiderlings eventually wander off to seek their fortunes.

➤ A ball of freshly hatched spiderlings.

Window spider *(Zygiella x-notata)*

Zygiella from Greek "join together"; *x-notata* from Latin "marked with an X"

The window spider is one of the most abundant European spiders, and it also occurs in North America. It can be seen almost anywhere there are window frames, at any time, day or night, summer or winter. My house has at least one of these spiders on every windowpane, sometimes one at each of the four corners.

Although this araneid can be readily identified by the gray leaf pattern on its back, it is the web and its position that give the game away. The webs of all *Zygiella* species are easily recognized because the upper part of the orb is free of spirals. When making the web, rather than moving round in a spiral, the spider reverses direction each time it reaches this segment. Directly behind the vacant sector is a tough strand of thread, the signal line that leads up to the spider's tubular retreat at the edge or corner of the frame. Here the spider bides its time with the tips of its front legs on the signal line, waiting for some insect to make contact. Sometimes spirals bridge the missing sector—the orbs of young ones, for instance, are complete—and in late summer the webs made by some adults may have the complete segment intact. However, there is something about the signal thread and the stretched oval hub that makes these webs unmistakable.

Window spiders are particularly active at night, when nocturnal insects such as moths are attracted to the light inside the house. Crane fly season, in late summer, seems a profitable time for these spiders, although the clumsy insects quite often manage to break

◄ Window spider with crane fly.

away before the spider has time to administer its subduing bite.

The eggs are laid in a silken cocoon that is often covered with a dense tangle of threads at the edge or corner of the window frame; it holds about 50 eggs.

Web of window spider. Note the vacant sector with signal line.

Orb weaver *Z. atrica*.

Orb weaver spider (*Zygiella atrica*)

Zygiella from Greek "join together"; *atrica* from Latin "of the house"

Almost identical in general appearance to the window spider is its close relative *Zygiella atrica*. The chief difference between the two lies in habitat: whereas the former chooses window frames around human habitations, *Z. atrica* prefers to build its web on heather, gorse and other bushes, often on open fields far from houses. This spider also thrives on rocks and breakwaters close to the ocean. The species name *atrica* seems rather inappropriate, as it rarely lives around houses; the name is perhaps better suited to its relative.

Unlike the window spider, which remains in its retreat when disturbed, *Z. atrica* will drop from its web like a stone, trailing a dragline that enables it to easily find its way back home. Although the two species are similar in appearance, they can be told apart, as *Z. atrica* has a more silvery folium and reddish marks on the front of its abdomen. This araneid is widespread throughout Europe but less common than its relative. It is also found in North America.

Bright warning coloration and spiny bodies deter predators.

Spiny-backed orb weaver
(*Gasteracantha cancriformis*)

Gasteracantha possibly from Greek "stomach" or "belly"; *cancriformis* from Latin "crab-shaped"

This odd-looking spider may not be a particularly large orb weaver but its combination of shape, color and "enamelized" abdomen makes it one of the most conspicuous spiders found in warmer regions of North America. Superficially it may resemble a crab spider, but *Gasteracantha* spiders have no relationship with the Thomisidae, being members of the Araneidae family.

The spiny-backed orb weaver—also known as the crab-like spiny orb weaver, among a range of English names—is the only species in its genus in the Western Hemisphere, where it can be found from the southern United States down to Argentina. Because of variations in its color pattern, over the years this spider has been described by numerous biologists under a plethora of names. It varies from whitish in Florida to orange-yellow in Central and South America, while the spurs can vary from black to red. Now it is considered a single species.

The female is wider than she is long, varying from 0.2 to 0.4 inch (5–9 mm) in length and 0.4 to 0.5 inch (10–13 mm) in width. The males are much smaller, longer than wider, and lacking the conspicuous abdominal spines. *Gasteracantha* is found in woodlands and citrus groves, where the webs are normally made between 3 and 20 feet (1–6 m) above the ground. It rests head down in the central disc, which is separated by an open zone from the sticky spirals beyond. When an insect is ensnared, the spider snaps the web around the capture area and rushes in to wrap the prey, then carries the victim to its headquarters on the central disc to be consumed.

Micrathena in its web showing stabilimentum.

Arrow-shaped thorn spider

(*Micrathena sagittata*)

Sagitta from Latin "arrow"; *micr* from the Latin "small" and *Athena* from the armor-wearing goddess who spun

Another enamelized araneid spider is *Micrathena*, of which several species are found in the Western Hemisphere. The one shown here is a female arrow-shaped *Micrathena sagittata*, being one of the species that thrives in the eastern United States and Central and South America. The female is the more spectacular of the sexes with her long, diverging abdominal spines, but like the previous species, *Gasteracantha*, she is prone to considerable variation.

The slightly slanted 12-inch (30 cm) orb web of *Micrathena* is constructed on low bushes around woodland edges, shrubby meadows and gardens, and is made with many radii and closely spaced spirals. Sometimes there is a small stabilimentum above the hub. Here the spider rests upside down on the web's downward slope. Like many spiders it will drop out of its web onto the ground when disturbed or when an intruder gets too close. It is reputed to prey largely on leafhoppers.

Rear view of a silver argiopid.

Silver argiope (*Argiope argentata*)

Argiope from the Greek name of a mythological nymph; *argentata* from Latin "silvery"

Argiopes are large, conspicuous araneids that hang head down in the center of their orb web. The web normally has a zigzag stabilimentum running through it that can come in a variety of forms; sometimes up to four are built in an X pattern, as is the case with this species. The purpose for these decorations is not certain. One reasonable theory is that the stabilimentum serves as camouflage, making the spider less conspicuous. This is supported by the fact that webs so endowed are made only by spiders that occupy the hub. An alternative explanation is that stabilimenta make the webs more visible so that birds are less likely to fly through them, saving the spider energy in repairs.

Argiopid spiders are largely tropical or semitropical species. The striking silver argiope, *A. argentata*, with its silvery cephalothorax, flared abdomen and yellow-and-black-banded legs, is found in the southern United States and Central and South America. Often several specimens may be found living in the same bush. Like a number of spiders, *A. argentata* does not necessarily bite its victim; its reaction to prey that become entangled in the web depends on the type of prey. The response to butterflies and moths is a long bite, whereas most other insects are first wrapped in silk. Presumably large insects, particularly those with loose scales, need to be rapidly subdued before they have a chance to escape.

➤ Silver argiopid in its web, showing a single stabilimentum.

Wasp spider, with yellow and black stripes.

Wasp spider (*Argiope bruennichi*)

Argiope from the Greek name of a mythological nymph; *bruennichi* after entomologist M.T. Brunnich

This exotic-looking creature is the sort of spider you would expect to find in a steamy rainforest rather than a rough clearing on the south coast of England. The female is not only large—especially in late summer, when full of eggs—but also boldly colored with transverse black and yellow wasp-like warning stripes. In contrast, the male is an insignificant little brown dwarf; indeed, it is difficult to imagine that the two are even related. Hardly surprising, therefore, that the female has little regard for her mate— she usually eats her suitor, sometimes doing so before mating has finished!

The large orb web, which is constructed near ground level by the female, in common with the rest of its argiopid kin, has stabilimenta running through it, one above and another below the hub. The webs of wasp spiders are inclined at an angle, usually in long grass around the edges of fields or wasteland, often where there is a good supply of grasshoppers.

In warmer countries there are many species of argiopid spiders, although the wasp spider, which is locally distributed in Europe, did not arrive in England till 1940, when it appeared on wasteland in Hampshire. Since then the spider has become well established on the south coast and is now extending its range northward. The one shown guarding her eggs in the huge flask-shaped egg-sac was photographed in Ashdown Forest in Sussex, well away from the coast. A similar species found in North America is the American garden spider, *Argiope aurantia*. Argiopid spiders are well represented in North America.

Close-up of long-jawed orb weaver showing its huge chelicerae.

Long-jawed orb-web weaver

(*Tetragnatha extensa*)

Tetragnatha from Greek "four-jawed"; *extensa* from Latin "stretched out"

Slim, tapering bodies, long, fine legs and outsized chelicerae usually distinguish the long-jawed Tetragnathidae spiders from other orb-weaving spiders, although not all species possess large jaws. They also have a habit of lying stretched out on a grass blade or twig, holding on with the third pair of legs and stretching the remaining three pairs in front and behind, parallel with their perch.

This cryptic pose, together with a body tinged with a light tracery of green and silver and dark veining, can make them very difficult to spot. The two commonest European species are *T. extensa* and *T. montana*, but they are tricky to tell apart in the field.

The webs of long-jawed spiders are delicate structures with few radii and widely spaced spirals; like *Meta*'s, they have a hole in the hub. The web is usually inclined at an angle or even horizontal. The name *Tetragnatha*—"four-jawed"—is fitting; in addition to the long, divergent chelicerae, their maxillae are similarly proportioned, giving the spider the appearance of having four jaws. These are used for mating, when the pair lock jaws in a wrestling posture, a precaution adopted by the male to prevent his being attacked by his mate.

The favorite haunt of long-jawed spiders is among reeds and rushes around the water's edge where there is a bounteous supply of flimsy, light-bodied insects such as gnats and mayflies to keep them happy. The webs are spun during the early evening before many of these insects become active. Long-jawed spiders, or grass spider species, as they are sometimes called, are common and widespread all over Europe as well as being found in North America.

➤ Long-jawed orb weaver with captured mayflies.

Long-jawed orb-web weaver in its web.

Long-jawed spider on grass blade.

Lesser garden spider on car side mirror.

Lesser garden spider

(*Metallina segmentata*)

Metallina from Greek "made of metal";
segmentata from Latin "flounced"

The lesser garden spider is the most abundant orb weaver found in northern Europe, but it is absent in North America. In some years, you only have to wander outside in the autumn in suitable places (which are all around) to see that almost every shrub and patch of long grass has at least one *Metallina* orb attached to it, with a yellowish spider decorated with a brown or purplish leaf pattern posing in the center. The markings are subject to much variation in depth and color. In the male these are an attractive rusty brown, and he has longer legs.

There are five European species in this genus of the Tetragnathidae, and their webs are relatively small for spiders of this size. Distinguishing characteristics of the webs are a small hole in the middle of the hub, closely set spirals and a greater number of radii than those of the closely related long-jawed spiders. As the spider does not hide away in a retreat, there is no signal thread; instead it rests in the middle of the web, ready to dash out at the slightest vibration.

During the mating season, in late summer and autumn, the males of the lesser garden spider can often be found hovering around the edge of the females' webs; sometimes the two coexist quite amicably in the web before courtship begins. The female triggers the actual mating by catching prey. While her chelicerae are busily engaged in wrapping or feeding, the male can approach safely and mate. This apparently amicable relationship is probably possible only because the male is as big as his partner. He also possesses much longer legs, an advantage if any sparring should develop. Such a relationship seems more amicable than is the case with most *Araneus* species. Female garden spiders, for example, are twice the size and more fearsome, and the male is lucky to get away with his life.

Although the lesser garden spider depends on its web for catching the vast majority of its prey, it has been known to fall vertically on a thread from the inclined web to seize some insect moving below, suggesting that there can be little wrong with its eyesight.

Cave spider with egg-sac.

Cave spider (*Meta menardi*)

Meta from Greek "cone"

The cave spider, *Meta menardi,* is a weird and rather large, creepy tetragnathid that spends its entire life in the pitch darkness (or near-darkness) of damp caves, sewers and the middle portions of long railway tunnels. As demonstrated by the female guarding her egg-sac in the photograph, the spider makes up for this with her subtle bronze and black markings, which are sometimes enhanced by a suffusion of yellowish patches, although these colors show up only when illuminated by a bright light.

Surprisingly, this spider is an orb weaver, but the web is difficult to discern in the gloom and is made even less visible by being constructed very close to the walls and roof of its drab habitat. Fortunately *M. menardi* is a very lethargic spider, which is just as well for arachnophobes who are able to pluck up sufficient courage to view this creature in its claustrophobia-inducing habitat!

M. menardi feeds on any invertebrates that venture into dark and damp places, such as woodlice, mosquitoes and hibernating butterflies and moths—I have found the remains of peacock butterflies and herald moths caught in their webs. Although by no means common, this spider is widespread throughout northern Europe and has also been recorded in some areas of North America. Be aware, though, that other species are also referred to as cave spiders, so it's best stick to scientific names whenever possible.

Golden orb spider (*Nephila clavipes*)

Nephila from Greek "fond of spinning"; *clavipes* from Latin "club-footed"

The largest araneomorph spider in the world is the golden orb spider, *Nephila clavipes*, a member of the Tetragnathidae. This impressive spider can be recognized not only by its size—the female has a body about 1.5 inch (3.75 cm) in length—but also by its long legs with conspicuous tufts of black hair on the femur and tibia of legs I, II and IV. The huge golden web, which can span gaps of up to 60 feet (18 m), also gives the game away. *Nephila* can be found in wooded areas from the southeastern United States through Central America and beyond. Unfortunately this spider does not grace the countryside of northern Europe.

I well remember the first time I ran into *Nephila,* during the early seventies, long before I had developed any particular interest in spiders. I was walking down a broad ride between pine trees in the Florida Everglades. I stopped in my tracks when I suddenly became aware of a sinister blurry form a few inches in front of my face, glinting in the late afternoon sun. Stepping back a few paces to focus, I became aware of the largest spider I had ever encountered, suspended in an orb web made from golden silk. The web was enormous—the orb itself was at least a yard (1 m) across and was held in place by thick silken strands that stretched across the 50-foot (15 m) wide track. Being a consummate arachnophobe at the time, I beat a hasty retreat.

Nephila sits in the center of her orb waiting for large insects—even the occasional hummingbird has been reported to have been taken. More often than not the male will be lurking close by in the web. In comparison with the female he is an insignificant creature that weighs about a hundred times less than his mate. Indeed, he is so small that the female ignores him as potential prey, even allowing him to crawl around her without fear of being eaten. He is also much more agile than she is, so he can easily escape from his relatively clumsy partner if the need should arise. It is interesting to note that whereas most male spiders have evolved a complex array of visual and tactile signals in order to herald their sexual intentions and avoid the disaster of becoming another meal, *Nephila* has developed this novel size-disparity approach to achieve the same end.

◄ Pair of golden orb spiders, the male is above.

➤ Golden orb spider showing femoral tufts.

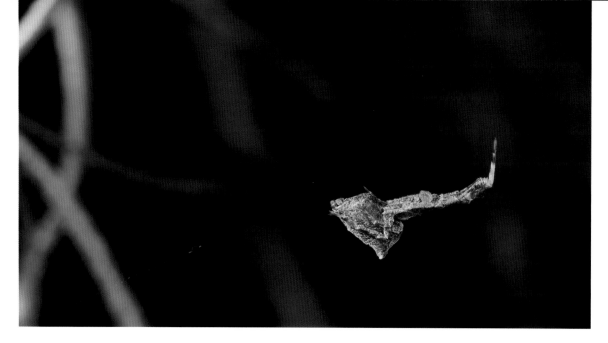

Lateral view of the garden center spider (*Uloborus plumipes*) showing cryptic attitude.

Garden center spider (*Uloborus plumipes*)

Uloborus from Greek "lethal"; *plumipes* from Latin "feathered feet"

The so-called garden center spider is a very recent addition to the fauna of Britain, where it can be found in increasing numbers in garden centers throughout the country. This member of the Uloboridae family is particularly common in the tropical humid biome of the Eden Project in Cornwall. *U. plumipes* builds a large, fragile-looking orb web among the plants or fabric of garden centers and greenhouses, thriving on the little flies and bugs found in such places, where insecticides are not routinely used.

This spider is unmistakable in appearance and capable of adopting a variety of strange poses, most often resembling a dried leaf or some other bit of debris. The best distinguishing features are the body shape and the tufts of hair on the front legs, together with its habit of sitting motionless with legs stretched out.

It is clear that this spider has been imported along with plants coming into the country from the European continent, as it has a wide distribution throughout much of Europe, the Mediterranean, Africa and a few areas in North America, where there are also a number of similar species.

◄ Ventral view of spider in its web.

7 Trappers: Disorderly Webs

The grouping of disorderly webs here does not always follow scientific rules, nor are the webs confined to specific spider families. Sometimes orb weavers spin untidy webs with little or no pattern—young cucumber spiders (*Araniella*), for instance—while some of the linyphiid sheet webs are far from being disorderly. Indeed, few webs are as disorderly as they may appear to us; nature always has sound reasons for adopting a particular design.

One of the scruffiest webs is produced by *Pholcus,* the daddy longlegs spider, and its appearance is not improved by the dust and debris that usually cling to it. The webs of dictinid spiders also appear disorderly. They are often made on dead or dying vegetation and consist of a dense weave of cribellate silk that is added to daily. The spider lives in the center, protected from would-be predators by the increasingly dense web.

The cobweb spiders, Theridiidae—or comb-footed spiders, as they are known in Britain—construct webs of irregular structure, but their varied designs are more precise than casual observation suggests. These are often called scaffold webs for reasons that become clear in the photograph on page 143. The central area of the snare is often a maze of threads, sometimes in the form of six-sided meshes or sometimes an open platform of loosely woven trelliswork. Their trapping action works as follows: sticky drops are placed along strategically placed threads, with the distribution depending on whether the web is intended to catch crawling or flying insects. Crawling insects form the principal prey of many species, including *Steotoda nobilis,* which constructs its trap attached to a wall or window or near the ground. When an insect blunders into a sticky blob at its base, the line ruptures and the insect is lifted into the air as the elastic thread contracts. This allows the spider to haul up the victim hand over hand. When it's in reach but at a safe distance, sticky threads are flung over the prey. Once the insect is sufficiently trussed up it is finally bitten on the leg. Theridiids will

◄ A balloning spider.

frequently tackle prey much larger than themselves, including potentially dangerous insects such as wasps.

Although theridiids have weak-looking chelicerae, they are compensated for by the powerful, rapidly acting venom. Unlike araneids, which chew their prey into pellets, the theridiids suck their victims dry, leaving a hollow husk behind.

When compared with orb webs, the sheet webs of linyphiids may seem disorderly, but those spun by many species are extremely clever, efficient and beautiful, especially when seen laden with fresh morning dew. Many have a scaffold-like superstructure and a platform beneath, shaped variously into concave or convex bows and domes. Insects knocked down by the trip wires above fall onto the web, while the spider lurking underneath strikes upward through the sheet.

Disorderly web-weaving families

The members of the **Theridiidae** family are commonly known as scaffold-web, cobweb or comb-footed spiders. Most spiders in this family are small to medium-sized and characterized by their glossy spherical abdomen and short legs. Although they have small chelicerae, their venom is potent, as evidenced by one of its members, the black widow (*Lactrodectus*).

Scaffold-web spiders are so named because of their web's scaffold-like structure (see page 138), while their alternative name, "comb-footed," stems from a row of curved serrated bristles on the tarsi of the hind pair of legs (which may be difficult or impossible to see unaided, particularly in adult males or the smaller

species). The comb plays a vital role in drawing out the viscous silk and flinging it over prey.

Some species are capable of producing audible mating calls by means of a stridulatory apparatus. This consists of a file on the rear of the carapace that is rubbed on a scraper equipped with teeth on the overhanging abdomen. North America has over 230 species, while northern Europe has 76.

The well-known **Pholcidae**, or daddy long-legs spiders, often construct their untidy or dome-shaped webs in houses. Others of the world's 870 species live in a wide variety of dark places, from inside caves to under logs. The typical pholcid can easily be recognized by its very long, slender legs and relatively small body, and its tendency to occupy dark corners. They also have an almost circular carapace with a characteristic eye arrangement. North America has 34 species, while northern Europe has three.

The **Dictynidae**, or mesh-web spiders, are a cribellate group of small spiders—they possess a cribellum and a calamistrum. The cribellum is a modified pair of spinnerets that produces thick, viscous silk that is combed out by a line of stiff hairs on the tarsus of the hind leg—the calamistrum—into a broad band of bluish silk. This traps prey by entanglement rather than by its adhesive properties. These organs are almost impossible to discern without considerable magnification.

The snares of dictynid spiders consist of an inconspicuous veil of threads radiating downward from the tops of low plants such as grass heads or bushes. Further lace-like cribellate threads are added to the radial ones, and it is these that catch the legs and wings of insects that are often considerably larger than the spider itself. The spider attacks its prey by

a breathing and circulatory apparatus that become less efficient with bulk.

Eleven similar species of so-called house spiders are found in northern Europe, but only two or three generally share people's homes, where they spend most of the time unobtrusively in their webs. Come autumn they may be spotted sprinting toward you across the carpet—these are most likely to be males of *Tegenaria domestica* looking for mates. Most other species live outside under stones, in tree trunks or amid thick vegetation. They are all generally smaller than *T. domestica*.

There is an exception to this size rule: an even scarier species, *T. parietina*. The female can grow up to 0.8 inch (20 mm), with exceptionally hairy legs spanning nearly 4 inches (10 cm). Not nice, but fortunately it rarely comes indoors, preferring to live outside in old buildings. Like most of its family, this spider builds large sheet webs extending from tubular retreats, usually in dark, dusty corners, where they lie in wait for prey.

Labyrinth spider immobilizing a butterfly.

Labyrinth spider (*Agelena labyrinthica*)

Agelena from Greek "gregarious"; *labyrinthica* from Greek "maze" (referring to the maze-like webs)

The labyrinth spider, like its close relative the European house spider, spends most of its life in or at the entrance of its tubular retreat, emerging onto the large expanse of sheet only to tackle prey or to mate.

The conspicuous webs are spun among low vegetation, usually grass or small shrubs, and are often present in large numbers during July and August. The web is a lot subtler than would appear at first sight. Although not sticky, the superstructure of scaffolding threads is very efficient at arresting flying or jumping insects and knocking them down onto the sheet below. The sheet's edges slope upward like a hammock, directing the prey into the center. What is extraordinary

is the way the spider itself can navigate through the maze of trip wires at lightning speed to deal with a captured insect, which staggers through them as though wading through molasses. After administering a few bites, *Agelena* drags the prey into the retreat to be eaten.

Mating takes place on the web, after which the male is allowed to remain with the female until he dies. Finally, in August, when ready to lay her eggs, she leaves her web to construct another home close by, which consists of a complex labyrinth of chambers. Here she lays her eggs and remains with the spiderlings until she too dies.

This spider is widespread in northern Europe, common in the south of England, but absent from Scotland. The ecological equivalent in North America is *Agelenopsis,* the grass spider.

➤ Web of labyrinth spider.

Lace-web spider

(*Amaurobius similis; A. ferox*)

Amaurobius from Greek "dark"; *similis* from
Latin "similar"; *ferox* from Latin "warlike"

For those who have not made the acquaintance
of *Amaurobius,* search around the nearest fence
or old wall and look for the characteristic untidy meshed web that surrounds its circular
retreat. When fresh, the web has a bluish,
lace-like appearance. Other favorite habitats
include gloomy places such as cellars, sheds,
crevices in bark and under stones and logs.

The best way of glimpsing this amaurobi-
diid spider or observing its hunting technique
is to venture out at night with a flashlight
and a tuning fork. Just touching the web with
the vibrating fork should initiate a rapid response from the animal lurking within, which
will shoot out to investigate the intruder and
maybe grab onto the fork with its fangs. The
large, thickset spider that emerges often has
dull, skull-like markings on the abdomen that
give the creature a rather sinister demeanor.

Although essentially nocturnal, this spider
will take prey at any time of the day or night,
hauling the victim into its den. The prey could
be any insect unfortunate enough to stumble
across the web, including flies, earwigs and
moths. I happened upon this one only because
I heard a wasp buzzing in its frantic efforts to
escape.

Besides their general appearance, *Amaurobius*
species can usually be identified by the cephalothorax, the dark head-end of which is raised
and shiny in contrast with the pale hind region. More specifically, if you are brave enough

◄   A large lace-web spider (*A. ferox*) ventures out of
its lair at night.

Lace-web spider (*A. similis*) retrieving a woodlouse.

to examine a specimen in your hand with the
aid of a magnifier, you will see that *Amaurobius*
possesses a cribellum just above the spinnerets
and a double-rowed calamistrum on the metatarsus of the hind leg.

Males are mature by late summer, when,
like the house spider, they may sometimes be
seen wandering away from their webs looking
for mates. There are five species of *Amaurobius* in
northern Europe, all broadly similar in appearance. *A. similis* is common, widespread and found
in North America, where there are also a number
of like species. *A. ferox* is a little larger and darker
than its relative and tends to prefer living under
logs and stones near walls and fences.

This tube-web spider is returning to its lair so rapidly that its body has twisted in two planes.

Tube-web spider (*Segestria florentina*)

Segestria from Latin *seges,* "cornfield";
florentina from Latin "from the city of Florence"

One of the most intimidating spiders likely to be encountered in northern Europe is a nocturnal member of the Segestriidae family, *Segestria florentina.* To start with, it's huge, having a body length of up to 1 inch (24 mm), several millimeters longer than the largest house spider.

S. *florentina* is the most impressive of three European species in this six-eyed genus of spiders. It lives in silk-lined tubes made in holes and crevices in old walls and trees; radiating from the entrance are about a dozen strong "fishing lines," as seen on page 160. One way to catch a glimpse of this formidable creature is to gently brush one of these lines with a blade of grass, upon which the occupant will dart out at awesome speed, flashing its green jaws, and bolt down its hole again. The action is so rapid that it is almost impossible to see exactly what the spider looks like. A fast flash photograph is one way to view it, but even the photo here—taken at one three-thousandth of a second—was not quite fast enough to arrest all movement.

A way to view this spider a little longer is to block the entrance hole with a stick before it returns, but extremely rapid reactions and strong nerves are needed to achieve this. If you are successful, this fierce spider will bite violently at the obstacle in its efforts to get back in. I don't mind admitting that my nerves

Fishing spider waiting on the surface of a pool, among sphagnum moss.

Raft or fishing spider (*Dolomedes fimbriatus*)

Dolomedes from Latin *dolo*, "pointed staff"; *fimbriatus* from Latin "fringed" (referring to the abdominal pattern)

A spider named *Dolomedes* must be an impressive one, and indeed this worldwide genus contains some species that are spectacular in both appearance and lifestyle. North America hosts several species, the best known being *Dolomedes triton,* while a very similar species, *D. fimbriatus,* is the European version. Like its relatives *Pisaura* and *Pisaurina* (page 56), *Dolomedes* is, along with other daylight hunting spiders, a nursery-web spider (Pisauridae family), but in view of its intriguing hunting methods it has been given special attention here.

Dolomedes is found in wet areas of acid bogs and heaths where there are permanent pools of water. With its cream-striped, dark chocolate-colored bulky body and thick legs, *D. fimbriatus* can appear quite awesome. The female specimen illustrated measured 1.04 inch (26 mm) in body length—probably one of the largest specimens ever recorded in England! It hunts by waiting motionless on the surface of the water, or on a natural raft consisting

of moss or a leaf at the water's edge, with its front legs resting on the water. By so doing it can detect the slightest ripple or vibration from an insect, tadpole or small fish moving beneath the surface; prey are then dragged out of the water and devoured. The spider has also been recorded tapping the water's surface with its toe to attract fish from beneath, and will often run across the water for some distance to pounce on an insect that has accidentally fallen in. When alarmed the spider will sometimes vanish by climbing down the stem of a water plant, where it can remain submerged for up to an hour.

Like her land-dwelling cousins, *Dolomedes* weaves a large silken tent around her egg-sac that she protects from nearby. On hatching, the spiderlings disperse, often moving away from the water to exploit higher, dryer places in surrounding shrubs.

There are two species of raft spiders in Europe, *D. fimbriatus* and *D. plantarius*, the latter being very rare and legally protected in the United Kingdom. Although widespread in Europe, raft spiders are very local in their distribution. In North America the genus occurs mainly in the east.

Water spider feeding on a waterlouse.

Water spider (*Argyroneta aquatica*)

Argyroneta from Greek "silver net"; *aquatica* from Latin "living in water"

The well-known water spider of the Argyronetidae family is the only spider in the world that spends almost its entire life underwater. Here, if you are lucky, you may see the spider pursuing aquatic creatures among the weeds, glittering like a ball of quicksilver.

All hunting takes place underwater, the spider deriving its oxygen from a bubble of air trapped among the hairs of its abdomen. To improve mobility the two pairs of hind legs are furnished with long, fine hairs so they act as paddles, a characteristic that makes it possible to identify even small, immature specimens.

◄ Water spider with bell.

This is a large species: males grow to more than 0.8 inch (20 mm), which is unusual among spiders, as males are almost always smaller than the females.

The water spider builds an underwater retreat held in place by a curved platform of silk that is then filled with air carried down from the surface—the whole structure resembles a sort of diving bell. Once it is filled, oxygen levels are maintained by oxygen from the water and bubbles from green plants diffusing into the bell, while carbon dioxide diffuses out into the surrounding water. Most activities take place within the bell, including mating, egg laying and consumption of prey. Young spiders do not make air bells but instead take over empty snail shells, which they fill with air.

Water spiders prefer living in still or slow-moving water and can be found all over Britain and northern Europe; they are very locally distributed but frequently abundant in those areas. This spider is absent from North America.

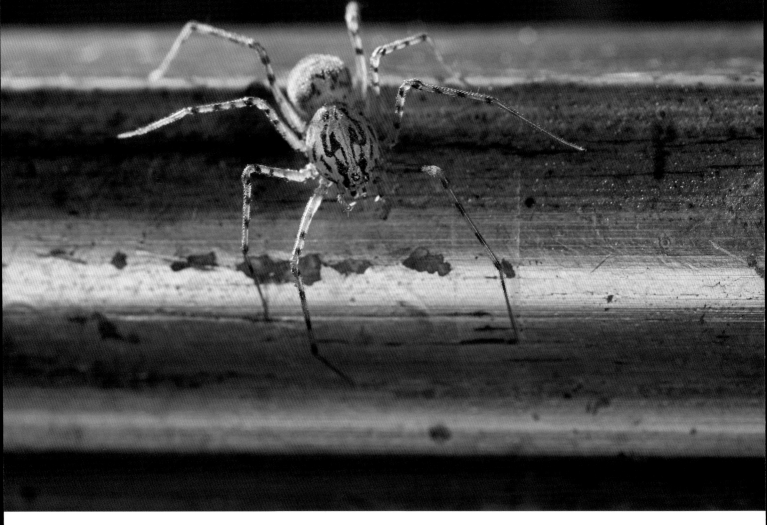

Spitting spider hunting on guilded picture frame.

Spitting spider (*Scytodes thoracica*)

Scytodes from Greek "marbled"; *thoracica* from Greek "chest"

This extraordinary spider was not discovered in Britain until 1816; during the next 120 years only six more specimens were found, all in southern counties. Since then the range of this scytodid has spread well into the Midlands and is probably increasing. It is almost certainly an import from the warmer regions of Europe; in Britain it can survive only inside houses, usually old ones, where it was probably introduced along with antique furniture. *Scytodes* can be found in many parts of the world, including North America, and as long as it is warm enough, at any time of the year.

The really astonishing thing about this spider is the unique way it catches and immobilizes prey. As twilight descends, *Scytodes* emerges from its daytime refuge behind picture frames and furniture to promenade the walls with its characteristic slow, measured gait. When it's about 0.4 inch (10 mm) from its prey, all you see is a quick jerk and then a struggling insect. On closer examination the victim can be seen to be tied down by a zigzag of 10 or 20 viscous threads on each side of its body, rather like Gulliver during his visit to Lilliput. The action is far too rapid to be seen by the unaided human eye, and even under a microscope the bonds are difficult to make out without special lighting.

The key to the mechanism that performs this trick is contained within the domed carapace, which houses two enormous glands, one loaded with poison and the other with a gummy substance. These are connected via ducts to holes at the tip of each fang. With a sudden contraction of muscles, the gum and venom are discharged under pressure through the fangs, which simultaneously oscillate rapidly from side to side, creating a controlled jet.

Scytodes can now deal with its victim at leisure without needing to wrap the body in a blanket of web. After a succession of small bites the struggling insect is subdued, allowing the spider to drag its prey free from the sticky bonds. Once digestive juices have been pumped into the insect, the body contents can be sucked out, leaving nothing but an undamaged empty husk behind.

The squirting technique can also be used for defense against spiders larger than itself, such as *Pholcus*. This unique ability, combined with *Scytodes'* capacity to move stealthily around the threads of larger and more aggressive spiders, ensures its safety in otherwise perilous situations.

When there is little food available in the winter, this spider remains concealed in cracks and crevices. It is long-lived, taking two or three years to become an adult, and with luck can eventually live to the ripe old age of four or five.

A spitting spider immobilizing prey on glazed glass.

Pirate spider (right) approaching a hammock-web spider.

Pirate spider (*Ero cambridgei*)

Ero from Greek "Eros," God of Love;
cambridgei from O. Prichard Cambridge,
19th-century arachnologist

Rather than stalking, chasing, ambushing or trapping prey like other spiders, pirate spiders lead a nomadic existence. These members of the Mimetidae family creep around stealthily rather like *Scytodes,* seeking out the webs of other spiders, especially those of theridiids, the ubiquitous *Achaearanea* (American house spider) being a favorite.

The little pirate will surreptitiously enter the tangled lines of the theridiid web and attract the attention of the occupant by plucking at the threads. Sensing potential prey or perhaps a mate, the web's owner investigates, and at exactly the right moment *Ero* grasps one of the theridiid's legs and bites the victim's femur. The effect of its extremely virulent venom—a venom specialized to kill spiders rapidly—is almost instantaneous, and the victor proceeds to suck the body juices out of the tiny hole it has made in the relatively large prey. On some occasions, though, the tables are turned and pirate becomes a victim of its own seduction.

Ero is the only genus of Mimetidae in northern Europe, this species being found amid low vegetation, trees and bushes. It spends its time wandering around in search of the webs of similar-sized spiders to invade. *E. cambridgei* is widespread and fairly common. Altogether there are four *Ero* species in northern Europe and more than a dozen in North America.

➤ This little pirate spider (*Ero* sp.) cannot really be appreciated unless magnified.

10 Photographing Spiders

This final section offers some basic help for those who may wish to try recording spiders using a camera. It assumes a rudimentary understanding of photography and photographic techniques, and many good books and a variety of field courses can provide both elementary and detailed guidance about all the aspects of photographic theory and wildlife photography.

Taking pictures of spiders is little different from photographing other small, active animals. You'll face the same challenges of exposure, focusing and sharpness. Fortunately, though, many of the technical obstacles that used to plague photographers, particularly those associated with close-up work, have been largely eliminated by modern equipment.

The first question you should ask yourself is what you are trying to achieve. You may wish to record spiders found on your travels, show spider behavior, create artistic masterpieces or perhaps a combination of all three. Knowing your goal will help with your general approach and the selection of suitable equipment for the task.

Film versus digital

Before launching into cameras, a word about digital photography. A few years back I would not have hesitated to say that film is the best medium in terms of image quality for wildlife photography, especially if medium format is used—it won hands down over digital. Now the tables have turned, and digital photography surpasses film not only in cost and convenience but, in my opinion, quality too. A few diehards still prefer film, and I can understand their emotional attachment to this great medium. However, while film may still have a few subtle advantages, they are hardly enough to justify struggling on with it any longer. I haven't exposed film since I tested the first truly high-resolution 35 mm digital camera (the Canon EOS 1DS) in

◄ Female wasp spider with cocoon.

2002—I was bowled over by the quality. More than 95 percent of the photographs in this book were taken using digital cameras.

The advantages of digital photography that are relevant to wildlife photography are summarized as follows.

Detail

A 10 megapixel or larger camera is generally capable of producing as much or more detail than the slowest, finest-grained 35 mm films available. Megapixels are not everything, though, as other aspects such as physical size of the chip, bit depth, lens design, camera circuitry and software also play crucial roles as a far as quality is concerned.

Noise and Grain

If noise can be kept under control, the image tones produced digitally are smoother than film, as are the transitions between tones, the best digital cameras being capable of generating less noise (graininess, in film terms) than film. This depends largely on the number of pixels and the relative size of the chip.

Speed of Results

One massive advantage of digital over film is that the image can be examined immediately after the picture has been taken, or, with cameras that support "live view" on the LCD, before the picture is exposed. Now we don't have to wait for days for the film to be processed to check any of the technical aspects of the image. Was the shot correctly exposed? Did the camera or spider move during the exposure? Have you focused in the correct place? All this and more is apparent immediately after the shot. The exposure can be checked in the field by examining the image on the camera's LCD monitor, or better still by referring to the histogram (a chart graphically depicting the tone levels from shadows to highlights). Composition and lighting too can be assessed on the LCD screen. If this is not sufficient—fine detail can be difficult to see in camera—the image can be transferred onto a computer where, with suitable software, all these factors can be carefully examined in minute detail.

File Types

The files generated by digital cameras come in two main types, JPEG and RAW, but a spider book is not the best place to explain their pros and cons. Suffice to say that if speed and minimum hassle are important, then JPEG files are the best option. If RAW files are used, they first have to be converted in a RAW conversion program to TIFF or JPEG files before they can be used effectively. The important point about RAW files is that during their conversion it is possible to make all manner of adjustments to image parameters such as exposure, contrast and color balance, with no or minimal loss in quality. This is a considerable advantage if your camera settings were lacking in precision or the finest quality is considered paramount.

Retouching and Spotting

Digital images are easily spotted and retouched using a computer. A word of warning, though—it is pointless to make adjustments to color, tone and contrast unless they are made on a good quality and regularly calibrated monitor. Otherwise you will be chasing your tail.

Index